Charles Babbage

and the Engines of Perfection

Owen Gingerich
General Editor

Charles Babbage

and the Engines of Perfection

Bruce Collier and James MacLachlan

Oxford University Press
New York • Oxford

Oxford University Press

Oxford New York
Athens Auckland Bangkok Bogotá Buenos Aires
Calcutta Cape Town Chennai Dar es Salaam
Delhi Florence Hong Kong Istanbul Karachi
Kuala Lumpur Madrid Melbourne Mexico City
Mumbai Nairobi Paris São Paulo Singapore
Taipei Tokyo Toronto Warsaw

and associated companies in
Berlin Ibadan

Copyright © 1998 by Bruce Collier and James MacLachlan
Published by Oxford University Press, Inc.,
198 Madison Avenue, New York, New York 10016

Oxford is a registered trademark of Oxford University Press

Design: Design Oasis
Layout: Leonard Levitsky
Picture research: Lisa Kirchner

Library of Congress Cataloging-in-Publication Data
Collier, Bruce.
    Charles Babbage and the engines of perfection / Bruce Collier and
James MacLachlan
        p.   cm. — (Oxford portraits in science)
    Includes bibliographical references and index.
    1. Babbage, Charles, 1791–1871—Juvenile literature.
    2. Mathematicians—England—Biography—Juvenile literature.
    3. Computers—History—Juvenile literature. [1. Babbage, Charles,
1791–1871. 2. Mathematicians.] I. MacLachlan, James H. 1928–   .
    II. Title. III. Series
QA29.B2C65   1998
510'.92—dc21   98-17054
[B]   CIP

ISBN 0-19-508997-9 (library ed.); 0-19-514287-X (paperback)

9 8 7 6 5 4 3 2

Printed in the United States of America
on acid-free paper

On the cover: The frontispiece of the October 1832–March 1833 issue of Mechanics
Magazine; inset: Babbage in 1860.
Frontispiece: Charles Babbage in 1829 as the Lucasian professor of mathematics at
Cambridge University.

Contents

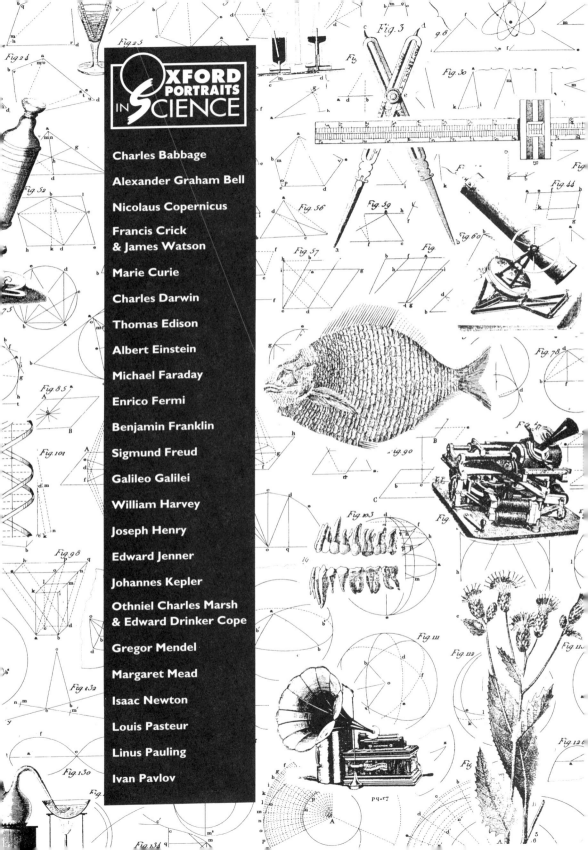

OXFORD PORTRAITS in SCIENCE

This watercolor miniature of Charles Babbage is one-half of a locket that also contains a portrait of his fiancée Georgiana Whitmore. The two were married in 1814.

The Making of a Mathematician

The two young friends were poring over columns of numbers. Two sets of clerks had calculated values for the positions of a number of stars as seen at regular times through the year. Now, the young men had to compare these results. As the number of errors mounted, they found the task increasingly tedious. Gentlemen of science, recent graduates of Cambridge University, Charles Babbage and John Herschel thought there had to be a better way.

"I wish to God these calculations could be done by a steam engine," Babbage complained. Herschel replied that he thought it might be possible. Babbage let the idea roll around in his mind for the next few days. Soon, he decided that not only was it possible, but he could do it.

This occurred late in 1821. By June of 1822, Babbage had constructed a small model of a calculating machine. He announced his success to the Royal Astronomical Society in London:

> I have contrived methods by which type shall be set up by the machine in the order determined by the calculation. The arrangements are such that . . . there shall not exist the possibility of error in any printed copy of tables computed by this engine.

Thus launched, Charles Babbage devoted many years of his long and productive life to the realization of his dream of mechanical calculations. Ultimately, his machine was abandoned. Although his principles were sound, the time and cost of construction proved greater than Babbage could afford. The government, which initially provided financial support, was unwilling to complete the project.

Charles Babbage was born in south London on December 26, 1791. His father, Benjamin, was a successful banker from Totnes in Devon (in southwest England). Benjamin had waited until he was 38 year of age and wealthy before marrying and moving to London to join a new banking firm. His wife, Elizabeth (Betty) Plumleigh Teape, was seven years his junior. Charles was born a year or so after their marriage. Later, two other sons died in infancy. A daughter, Mary Anne, was born in 1798. She outlived Charles and the two siblings remained close throughout their lives.

As a child, Charles displayed a great curiosity about how things worked. With each new toy, he would inquire, "Mamma, what is inside of it?" Often, if he was not satisfied with the answer, he would break open the toy to see for himself. Once, his mother took him to see an exhibition of machinery in London. Charles showed so much interest in one exhibit that the artisan took him to his workshop. There, the boy was fascinated to see a foot-high silver figurine dancing on a stand and holding a bird that flapped its wings and opened its beak. Though Charles was curious about the mechanism within, he did not break open this toy. However, many years later, he purchased the figurine at an auction. He restored it to working order and proudly demonstrated its antics in his drawing room.

At age ten, Charles suffered from violent fevers. In that time before modern drugs and innoculations, his parents feared for his life. Hoping that country living would improve his health, they sent him to a school in Devon near

Totnes. The schoolmaster was asked to attend to his health, but not to press too much knowledge on him. In later life, Babbage wrote that this mission was "faithfully accomplished. Perhaps great idleness may have led to some of my childish reasonings." One of his childish reasonings involved performing experiments to see if devil-worship incantations actually worked. For him, at least, they did not.

By 1803, Benjamin Babbage had amassed sufficient capital to retire. With his wife and daughter, he returned to Totnes. At the same time, in improved health, Charles was sent to a small residential school in the village of Enfield near London, where he remained for three years. The teacher at Enfield was Stephen Freeman, an amateur astronomer. He awakened Charles's interest in science and mathematics. Yet Babbage's mathematical skills were largely self-taught from books he found in the school's modest library. In his second year at Enfield, Charles and another boy began getting up every day at 3 A.M. to study algebra. When Freeman learned of this several months later, he made them stop. However, Babbage thought highly enough of Freeman's school that he later sent two of his own sons there for a time.

Charles then moved to a small school near Cambridge for a couple of years. This may have been to prepare for entrance to the University of Cambridge, but it made little impression on him. At age 16 or 17, Babbage returned to Devon to live with his parents. He learned Latin and Greek with a tutor and also spent much time studying mathematics on his own. By then, he was passionately fond of algebra and devoured every book he could find on the subject.

Trinity College, Cambridge, was founded in 1546. This was the college of Isaac Newton and Charles Babbage, both of whom also held the Lucasian chair of mathematics at Cambridge.

In the fall of 1810, Charles Babbage enrolled at Trinity College, Cambridge. This was the university of Isaac Newton, inventor of calculus and the theory of gravitation. Babbage looked forward to receiving a first-rate training in mathematics, but was destined to be greatly disappointed. For a century after Newton's tenure, Cambridge had advanced very little beyond him in the study of mathematics. In fact, almost all advances since Newton had been made by French and Swiss mathematicians. These men followed a style of calculus invented about the same time as Newton's by a German, Gottfried Leibniz. Although the two had invented the calculus independently, the English claimed Leibniz had stolen his ideas from Newton.

Sir Godfrey Kneller, the most popular portrait painter of his time, produced the first portrait of Sir Isaac Newton in 1689, when Newton was 46.

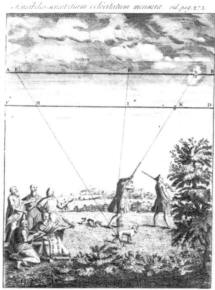

THE

METHOD of FLUXIONS

AND

INFINITE SERIES;

WITH ITS

Application to the Geometry of CURVE-LINES.

By the INVENTOR

Sir I S A A C N E W T O N, *K.*
Late Prefident of the Royal Society.

Tranflated from the AUTHOR's LATIN ORIGINAL
not yet made publick.

To which is fubjoin'd,

A PERPETUAL COMMENT upon the whole Work,

Confifting of

ANNOTATIONS, ILLUSTRATIONS, and SUPPLEMENTS,

In order to make this Treatife

A compleat Inftitution for the ufe of LEARNERS.

By *JOHN COLSON*, M.A. and F.R.S.
Mafter of Sir *Jofeph Williamfon*'s free Mathematical-School at *Rochefter.*

LONDON:
Printed by HENRY WOODFALL;
And Sold by JOHN NOURSE, at the *Lamb* without *Temple-Bar.*
M.DCC.XXXVI.

Calculus provides a way to calculate changing quantities, for example, to find the changing speed of a jet of water from a hole in a barrel as the water level in the barrel decreases. Newton thought of the quantities as being *in flux*, and called his technique the study of *fluxions*. Leibniz, on the other hand, thought of the successive *differences* as a quantity changed, and called his technique the study of *differentials*. Also, the two men differed in the way they symbolized the changing quantities; that is, they had different mathematical notations.

Babbage was keen to be up-to-date in mathematics when he got to Cambridge. Having an annual allowance from his father of £300, Charles decided that, on his way from Devon to Cambridge, he could stop in London and splurge on the best calculus textbook available, which was a three-volume work by the French mathematician Sylvestre-François Lacroix. He expected it to cost £2 (about a third of a week's allowance) but discovered that

The Method of Fluxions and Infinite Series is one of three mathematical works by Newton that are the basis for the historical claims of his priority over Leibniz as the inventor of calculus.

Gottfried W. Leibniz, philosopher, mathematician, and historian, was also a member of the royal court at the house of Hanover in Germany. When the Elector of Hanover became King George I of England, Newton (Leibniz's arch-rival) persuaded the king not to bring Leibniz to London.

England's war with Napoleon had driven up the price of French books. So, he paid out the £7 that the bookseller demanded. He would just have to buy less wine for a few weeks.

Once settled in Cambridge, Charles plunged into his new book. Soon, he ran into some mathematical reasoning he could not understand. He took the problem to his assigned tutor, John Hudson. After listening to the question, Hudson told Babbage that such a question would never be asked on any of his university exams, and he would do better to spend his time on the kinds of questions that would.

Another Cambridge tutor, Robert Woodhouse, had written books on the newer style of mathematics, but they had little influence. An English review of one of Woodhouse's books criticized it unmercifully:

Mr. Woodhouse's quitting the fluxionary notation of Sir Isaac Newton for the differential one of Leibniz, who, though a man of eminent and diversified talents, was certainly a plagiarist in matters of science, strikes us as a ridiculous piece of affectation. The two calculuses differ only in name and in notation, which in fluxions is equal, at least in simplicity to that of differentials, and unquestionably superior to it in point of conciseness. As this is the case, and as the Royal Society of London took a great deal of pains to have Sir Isaac's claim to the invention investigated and established, we trust the principal mathematicians in this island will never think of abandoning the notation of the inventor for the other.

This came 90 years after the dispute between Newton and Leibniz. It neglected to mention that Newton himself had written the indictment of Leibniz's calculus!

Babbage quickly realized that, if he wanted to become a mathematician, he would have to continue to study on his own. He would get no help from his teachers. Evidently, the Cambridge faculty were so dazzled by Newton's achievements that they felt incapable of surpassing them in any respect even though Cambridge prided itself on the quality of its mathematics education. Indeed, all England recognized a Cambridge degree in mathematics as the unexcelled preparation for professional life, whether in law, medicine, or theology. Yet, the examinations did not test mathematical competence as much as they did the students' capacity to memorize set pieces taken from the works of Newton. As far as Babbage could see, they were a hundred years out of date.

It was not long before Babbage decided he had to do something about that. During his second year at Cambridge, Babbage jokingly suggested to a friend that they should have a society to promote Lacroix's textbook among their fellow students. This was because another student group had just been formed to promote the reading of the Bible. Babbage drew up a small poster on behalf of

Sir John Frederick William Herschel, a life-long friend of Babbage, was an astronomer like his father, who discovered the planet Uranus. Besides creating a map of the southern sky from Cape Province in Africa, John Herschel was also a pioneer in photography.

Lacroix's book as a parody of the posters the Bible Society had plastered around Cambridge. But his friend took him seriously, and a few days later, a dozen students met to found the Analytical Society.

The Analytical Society held monthly meetings during school terms from 1812 to 1814. Some of the society's work was published in a small book in 1813. However, its most productive result was the publication of two books on the calculus of differentials. The first was a translation of part of Lacroix's work by Babbage and two friends that appeared in 1816. Four years later, the same three men produced a two-volume set of examples of problems in the calculus. Babbage's two friends were John Herschel and George Peacock.

Herschel was the son of William Herschel, the outstanding astronomer who had discovered the planet Uranus in 1781. John followed in his father's footseteps, and became one of the leading men of science in England during the 1800s. He and Babbage became lifelong friends, and Babbage named his firstborn son Herschel. George Peacock spent most of the rest of his life at Cambridge as a mathematics tutor. He became a force in reforming the mathematics curriculum there, although it took many years to accomplish.

Babbage made other lifelong friends among the members of the Analytical Society. One was Edward Bromhead, after whom Charles named another son. Bromhead inherited his father's estates in Lincolnshire, and spent most of his life managing them. Another friend was Edward Ryan, who became chief justice for the state of Bengal in India.

You should not suppose that Babbage spent all his time on mathematics. He was, in fact, a popular and gregarious student, with friends of widely ranging interests. He met one group for breakfast every Sunday morning to discuss many philosophical issues, such as the meaning of life and death. With another group, he often sailed on the river Cam in his own boat. These friends were chosen not for their intellect but for their ability to row the boat when the wind dropped. Babbage was also a keen player of table games—chess and whist, which is a card game like bridge. Babbage was also interested in chemistry. He set up one of his rooms as a laboratory, where he conducted experiments, often assisted by John Herschel.

To get some idea of Babbage's lifestyle, one needs to convert the currency of his time into present-day values. For a rough comparison, consider that £1 (one pound sterling) in the early 1800s is equivalent to about $200 at the end of the 1900s. So Babbage's allowance of £6 per week would represent about $1200 today—not too shabby. Of course, prices then were not the same as today. Generally, manufactured goods were more expensive; the necessities of life were cheaper. The wage of an ordinary clerk or laborer in England at that time was about £1 per week. These working poor managed to raise their families on such an income. Commodity prices were so low that £1 would buy 50 pounds of meat.

Babbage's weekly expenditure might well have been greater than £6, because he spent his summers at home in Devon. Presumably, his father did not charge him room and board then. In the summer of 1812, Babbage and his friend Edward Ryan met the two youngest of the eight Whitmore sisters, whose home was in Shropshire. Romance blossomed, and before the summer ended, Charles was engaged to Georgiana Whitmore, who was just a year younger than he. Ryan became engaged to her sister, Louisa.

For many Cambridge students, the most important activity was preparing for the examinations. Obtaining high honors was the surest way to gain good employment. A student guide of the period advised that having numerous friends was the best way to waste time. It also deplored as "the first step to idleness and folly, the reading of books you think are suitable instead of those recommended by your tutor." This was advice Charles Babbage did not follow. According to one of his tutors, Charles did not care to be ranked and wished only for his tutors to be aware that he knew the work. Moreover, this tutor remarked, "he would not compete for mathematics honors on taking his degree, though I believe that if he had, he could easily have taken first place." The summer after he graduated, Charles wrote to John Herschel a direct contradiction of the student guide's advice:

> There are two reasons for which I shall always value a university education—the means it supplied of procuring access to books—and the still more valuable opportunities it affords of acquiring friends. In this latter, I have been singularly fortunate. The friendships I have formed while

there I shall ever value; nor do I consider my acquaintance with yourself as one of the least advantages.

Babbage graduated at Cambridge in the spring of 1814. Against his father's wishes, Charles married Georgiana Whitmore in July. Benjamin Babbage had no complaints against Georgiana. His attitude was that, like himself, Charles should wait until he was properly established financially. The young couple honeymooned in a charming village in Devon. From there, Babbage wrote a letter describing his situation to John Herschel, and then went on to include some mathematical theorems he had been working on. Herschel was appalled. He replied to Charles: "'I am married and have quarreled with my father'—Good God Babbage—how is it possible for a man calmly to sit down and pen those two sentences—and then to pass on to functional equations?"

Georgiana Whitmore married Charles Babbage in 1814 while Charles was still an undergraduate at Cambridge University.

The newlyweds spent a long romantic summer in the Devon countryside. In the fall, they moved to London. Despite his father's urgings, Charles had no job and few prospects. Fortunately, Benjamin continued his £300 annual allowance, to which Georgiana could add £150 of her own. With such an income, the couple could maintain a modest life without lavish entertaining.

2

In Scientific Circles

Charles and Georgiana Babbage moved to London in the autumn of 1814. After a few months in various quarters, they moved into a small, comfortable house in the Marylebone district just south of Regent's Park in London's northwest. The previous month, on August 6, 1815, Georgiana had given birth to Benjamin Herschel Babbage, who was always called by his second name. Other children were born at approximately two-year intervals: Charles Jr., Georgiana, two sons who did not survive infancy, Dugald Bromhead, and Henry Prevost.

These early years in London were generally happy. The Babbages often visited with friends and relatives in other parts of England. Normally, they spent the summer months in Devon, with side trips to Shropshire to visit the Whitmores. Charles was a somewhat grim and distant father, though he tried to overcome his experiences with his own father. He described his father to his friend John Herschel in a letter:

> He is stern, inflexible and reserved, perfectly just, sometimes liberal, never generous. [He has] a temper the most horrible that can be conceived. A tyrant in his family, his

In 1814, newlyweds Charles and Georgianna Babbage moved to the Marylebone district, just south of London's fashionable Regent's Park. Here, ice skaters take advantage of the park's frozen pond in the 1830s.

presence occasions silence and gloom. . . . Tormenting himself and all connected with him, he deserves to be miserable. Can such a man be loved? It is *impossible.*

This was Charles's model for fatherhood. Perhaps he may be forgiven if he sometimes fell short of the higher expectations he tried to fulfill for himself as a father. As children, the two younger boys were in considerable awe of their father; however, in later life, Henry was his father's friendly assistant for a time.

For a while, Charles sought paid employment, to prove to his father that he could make something of himself. In 1816, he applied for the post of math professor at a college a few miles north of London. It paid a salary of £500. He had strong letters of recommendation from two outstanding men. However, he was told that he would not get the job because he lacked influence with the board of directors. Three years later, again with strong recommendations from eminent mathematicians, he missed a post in Edinburgh because that job went to a Scot. Indeed, Babbage's spirit of independence would not make it easy for him to gain any employment.

With a barely comfortable income from their parents, Charles and Georgiana managed. Charles continued to work on the mathematical topics he had studied in Cambridge. In addition, he set up a workshop in one of his rooms to explore interesting experiments in chemistry and mechanics. Also, he began to make himself known to the scientific bright lights in London. John Herschel lived nearby and introduced Babbage into scientific circles. The Herschels, father and son, sponsored Charles's membership in the Royal Society. Founded in 1662, the Royal Society was England's major scientific insitution. Isaac Newton had been its president from 1702 to 1727.

The Royal Society published a monthly journal of scientific papers. From time to time it also supported scientific expeditions abroad. Charles published a 111-page essay on calculus in the *Philosophical Transactions of the Royal Society* in

1815–16. Also through the Herschels' influence, Charles was asked to present a series of lectures to the Royal Institution in London in 1816. Founded in 1800, the Royal Institution was both a research lab and a public forum for science. Its director, Humphrey Davy, conducted important chemical research and discovered several new elements. His successor, Michael Faraday, would later do important work in electromagnetism. Both those men gave outstanding popular lectures on science to the cream of London society. Charles's lecture series was well received. It demonstrated his capacities as a scientist and put him into the center of London society, both scientific and otherwise.

Besides social visits with his family, Charles Babbage also frequently traveled abroad for scientific purposes. In 1819, he and John Herschel went to Paris to visit its eminent

Pierre S. La Place, a French mathematician, has been called the Isaac Newton of France. He wrote an important work on celestial mechanics, and also helped to found studies of probability theory and thermochemistry.

scientists. Among others, they met and became friendly with Pierre Laplace, Claude Berthollet, Jean Fourier, Jean Biot, and François Arago. Laplace was a theoretical astronomer who did much to extend and deepen Newton's analysis of the planetary system. Laplace had also held high office in Napoleon's government. Babbage remarked that no scientist would expect to achieve that status in England. Berthollet, an eminent chemist, was active in the improvement of industrial processes such as the dyeing of fabrics. Fourier was an outstanding mathematical physicist. Babbage recorded that "his unaffected and genial manner, and his admirable taste conspicuous even in his apartments, were most felt by those who were honored by his friendship."

Jean B. Biot, a French physicist, studied polarization of light, the magnetic effects of electricity, and the flow of heat in solids.

Biot was a balloonist, and an active investigator of phenonomena of light, electricity, and magnetism. Late in Biot's life, Babbage visited him, inquiring of a servant if his health could stand the visit. From his bedroom, Biot heard the remark and came out into the hall saying, "My friend, I would see you even if I were dying." The physicist François Arago was a co-worker of Biot's, and also active in the governments of his day. His work was recognized by the Royal Society of London, which gave him its Copley Medal in 1825.

Babbage and Herschel returned to London full of admiration for the way science was organized in France and impressed by the influence scientists had with their government. They felt there was a lot of room for improvement in England. One result of those feelings developed early in 1820. The two young men were discontented with the state of the Royal Society. It seemed to them to be much more a high-prestige social club than a real scientific society. Only about a third of its members actually had any scientific training. Realizing that the Royal Society was contributing little to astronomy, they resolved to form a society of astronomers. Together with Francis Baily and eleven others, on Wednesday January 12, 1820, they dined at the Freemason's Tavern in London to organize the Astronomical Society of London.

Their friend Francis Baily was an interesting character in his own right. His banker father apprenticed him to a firm of merchants in London in 1788. In 1798, he joined a firm of stockbrokers and amassed a considerable fortune. Around 1810, he spent some time investigating interest rates for life-annuity investments. When his leisure time increased, Baily took up the study of astronomy. With his mathematical training and interests, he later engaged in producing accurate tables of star positions to supplement the *Nautical Almanac*, which was a then-inaccurate government publication intended for navigational use at sea. In 1836, Baily made close observations of an eclipse of the sun. He reported a series of bright spots along the rim of the moon-sun boundary just before totality. The phenomenon is named "Baily's beads" in his honor.

Baily became secretary of the new Astronomical Society, with both Babbage and Herschel as members of its first executive board. To enhance the society's prestige, the board members sought as president Edward Seymour Duke of Somerset, who had been president of the Royal Institution. Babbage was friendly with the Seymour family, which had estates near Totnes in Devon. However, the Duke was also a good friend of Sir Joseph Banks, president of the Royal Society for more than 40 years. Banks jealously protected the Royal Society's influence and vigorously opposed any steps that seemed to threaten his power. Banks persuaded Somerset to decline the presidency of the Astronomical Society. The board then approached Sir

Crane Court, the first permanent home of the Royal Society, was purchased in 1710.

William Herschel, who agreed to let his name stand as long as he was given no duties. Banks died in 1820 and was replaced by Sir Humphrey Davy as president of the Royal Society. The general situation of science in England changed very little under Davy's rule, so that, ten years later, Babbage would mount a stronger challenge to the monopoly of the Royal Society.

Once started, the Astronomical Society energetically pursued the improvement of astronomy in England. In particular, it was active in enlarging and correcting the tables in the *Nautical Almanac*. This was an effort that would take more than 15 years to accomplish. The Astronomical Society thrived, and received its Royal Charter in 1830, when it had attained a membership of 250. A historian of the Royal Society notes that Banks's fear that the competition of new societies would be detrimental to the Royal Society was without foundation; instead, their contributions to research "have greatly promoted the advance of science and have raised its standing in this country."

In 1821, the Astronomical Society assigned Babbage and Herschel one of the tasks for improving the tables of the *Nautical Almanac*. They constructed the appropriate formulas and assigned the arithmetic to clerks. To diminish errors, they had the calculations performed twice, each by a different clerk. Then they compared the two sets for discrepancies. Of course, none were apparent if both clerks made the same error, but it was better than having the two mathematicians do all the routine arithmetic—and they could make errors too.

It was during the course of this activity that Charles Babbage began to seriously consider how such routine calculations could be performed mechanically. In the following months, he made several designs for clockwork-like mechanisms that could be made to control a set of wheels with numbers along their edges that could print on paper. Details of the design of Babbage's machine, his *Difference Engine*, are discussed in the next chapter.

By the end of the spring of 1822, Babbage had constructed a small Difference Engine that would produce six-place numbers. Unlike most men of science at the time, Babbage had a small lathe in his workshop. However, it was not elaborate enough to produce the accurate wheels he needed. So he had them turned and ground at a professional machine shop. He built the frame himself and mounted the axles and wheels.

In June 1822, Babbage was secure enough about his machine and its principles of operation to announce it publicly at an Astronomical Society meeting. He also wrote an open letter to Sir Humphrey Davy describing the Difference Engine in considerable detail. Babbage had this letter printed and distributed around London. When the letter came to the attention of the British government, it asked the Royal Society to judge the worth of the invention. Replying promptly on May 1, 1823, the Royal Society members reported that "they consider Mr. Babbage as highly deserving of public encouragement in the prosecution of his arduous undertaking." His own Astronomical Society was so impressed that it awarded him its first gold medal in 1824.

The British government advanced Babbage a fee of £1500, and he began to construct the full Difference Engine, which would require about 20 sets of wheels, all interacting with great precision. Babbage needed a small factory and competent workers. To that end, he sought advice from a fellow member of the Royal Society, the engineer Marc Isambard Brunel.

Marc Brunel, born and trained in France, was a civil engineer. For a while in the 1790s, he was chief engineer for the city of New York. Then, in 1799, he sailed for England with a great idea. He had designed machinery that would mass-produce pulley blocks for sailing ships. A naval warship was equipped with 1400 of these blocks, which until then had been made by hand one at a time. Brunel engaged the London machinist Henry Maudslay to build the machinery he had designed. With 43 machines for cutting and shaping the wooden and metal parts, ten men could produce as many blocks (of superior quality) as 100 men had previously made with hand tools.

In 1814, Brunel was elected to the Royal Society, where he became friendly with Charles Babbage. In 1823,

Brunel recommended to Babbage that he hire one of Maudsley's workmen to construct the Difference Engine. Maudslay was renowned for the high precision of the machine tools he produced. His employee, Joseph Clement, would be just what Babbage needed. Charles converted three rooms in his house into a workshop, with a forge in one of them. Clement started with one lathe in his own kitchen. Soon, with funding from both Babbage and the government, Clement greatly expanded his workshop. For eight years, parts for the Difference Engine flowed back and forth between the two establishments. Babbage conducted trials and experiments, while Clement fabricated the parts. At the same time, Clement built up the number and quality of his machines and his mechanics. One of Clement's mechanics was Joseph Whitworth, who later became the leading manufacturer of precision machinery in England.

As Babbage delved more deeply into machinery, he realized there was a lot he could learn from other artisans. Soon, he was touring craft and manufacturing establishments all over England and in Scotland. Sometimes Georgiana accompanied him, making a holiday of the trip. On several occasions, Babbage took along the young son of the Duke of Somerset. Through these trips, Charles gained considerable knowledge of British industrial practices. He was often consulted by friends interested in investing in such enterprises. Had it not been for his obsession with calculating engines and his spirit of independence, he might have become an outstanding consulting engineer. However, besides calculating machinery, there was no other area to which he would devote his full attention.

Once the construction of the Difference Engine was underway, Babbage did make occasional forays into other fields. In 1824, with Francis Baily's influence, Charles was invited by some investors to organize a life insurance company. The new challenge intrigued him, and he threw himself into the task of determining the appropriate rates to charge for life insurance policies. This required him to investigate age-dependent death rates (actuarial tables) and rates of interest on invested funds. As it happened, the project fell through when several of the investors backed out.

Having collected so much information, Babbage decided that he would have to make some other use of it. In 1826, he published a book on the life insurance industry, *A Comparative View of the Various Institutions for the Assurance of Lives*. In fewer than 200 pages, this book provided a very useful consumer's guide to the life insurance companies in England at that time. Readers could use it to compare companies and make intelligent decisions about which one would suit their particular needs.

In the process of designing and building his Difference Engine, Babbage required many accurate drawings of the parts. While using these drawings, he felt that they did not fully and adequately describe the mechanism. For a machine with many parts moving in various ways, the static drawings could only show the shape and arrangement of the parts. So Charles devised a system of *mechanical notation* that would also indicate how the parts moved—their speeds and interconnections. Unlike the usual drawings, the notation did not picture the shapes of the parts. Rather, it was a table of numbers, lines, and symbols to describe the machine's actions. It was a general system that could be used to describe any machine. Perhaps the simplest comparison you can make is to musical notation. Violinists who can read sheet music are able to translate sharps, flats, and eighth notes into how to place their fingers on the strings and how to move the bow. In the same way, a mechanic who understood Babbage's notation would be able to translate it into an understanding of a machine's operations. Charles published a description of his mechanical notation in the *Philosophical Transactions of the Royal Society* in 1826. However, this mechanical notation did not ever come into widespread use.

At the same time that Charles continued to direct the construction of the Difference Engine, he also investigated existing tables that are important in calculations. Before the advent of electronic calculators, the multiplication of large numbers was performed using tables of logarithms. Logarithms are based on the idea in algebra that powers are multiplied by adding their exponents (or indices); for example, $n^a \times n^b = n^{a+b}$. For most calculations, n represents 10, and formulas are used to make tables of exponents (or logarithms) that represent the numbers you wish to multiply. For example, $2 = 10^{0.30103}$, $3 = 10^{0.47712}$, and $6 = 10^{0.77815}$. That is,

Number	Logarithm
2	0.30103
3	0.47712
6	0.77815

Notice that the sum of the logarithms of 2 and 3 is the logarithm of 6:

since $2 \times 3 = 6$, then $\log 2 + \log 3 = \log 6$.

With a table of logarithms, if you wish to multiply two large numbers, you need only add their logarithms. This makes calculations simpler and much quicker. But someone has to construct the table first.

The very first table of logarithms had been published in England 200 years earlier. Babbage compared several tables published since then. Wherever they differed, he recalculated the value so that he could produce a table completely free from error. With the help of an army engineer, he directed the work of a number of clerks. The corrected table was published in 1827. This table was reprinted many times, even after 1900.

In February of 1827, Charles's father died in Devon at the age of 73. Old Benjamin left sufficient funds to care for his wife, Betty, who moved to London to live with Charles

text continues on page 33

LOGARITHMS EXPLAINED

ogarithms come from the mathematical operation of exponentiation. Multiplication means adding a number to itself some number of times. Exponentiation means multiplying a number by itself some number of times. Consider the following:

10 to the "zeroth" power (10^0) is, by convention, 1.

10 to the 1st power (10^1) is ten itself.

10^2 (ten squared) is 10×10, or 100.

10^3 (ten cubed) is $10 \times 10 \times 10$, or 1,000.

Fractional exponents are also possible. Thus, $10^{0.5}$ (the square root of 10) is the number that yields 10 when multiplied by itself. Because $3 \times 3 = 9$ and $4 \times 4 = 16$, you can tell that $10^{0.5}$ will be somewhere in between. It is, in fact, about 3.162.

In general, you can produce any desired number by raising 10 to some power. Thus, we can get Babbage's year of birth with $10^{3.2531} = 1791$. Now, taking the logarithm (abbreviated log) of a number involves posing the question the other way: "What power would I raise 10 to in order to get this result?" For the number 1791, the answer is 3.2531. This can be written:

log (1791) = 3.2531

This is not useful yet, but it becomes so with a few more facts. Consider any two numbers, called A and B. Then

$\log (A \times B) = \log (A) + \log (B)$

$\log (A \div B) = \log (A) - \log (B)$

$\log (A^B) = \log(A) \times B$

That is, working with logs rather than the raw numbers allows us to substitute addition for multiplication, subtraction for division, and multiplication for exponentiation; and in each case, the first operation is much easier to perform by hand than the second.

Suppose, for some odd reason, you wanted to raise the number of children born to Charles and Georgiana Babbage (8) to the power of his age when they got married (22.5) to get $8^{22.5}$. You could multiply 8 by itself 22.5 times, if you had the patience, but it would take a long time. Or you could use logs:

log (8) = 0.90309

$0.90309 \times 22.5 = 20.319525$

Now, you know that the log of your answer is 20.319525. To find that answer itself, you need to consult your log table to find the antilog of 20.319525, the number equal to $10^{20.319525}$. The answer is approximately 208,701,000,000,000,000,000.

text continued from page 31

and his family. Charles inherited an estate worth £100,000. The interest on the investments and the rent on the properties provided a comfortable income for the rest of his life. However, his view of a comfortable life did not last long. In July of the same year, Charles Jr. was struck with a childhood disease and died at the age of 10. Then, less than a month later, Charles's wife Georgiana contracted a serious illness. At the end of August, both she and a newborn son also died.

Charles was devastated.

His mother, Betty, was able to look after the remaining three sons and one daughter. Charles sought solace at the home of his friend John Herschel and his family. Betty wrote to Herschel in early September: "You give me great comfort in respect to my son's bodily health. I cannot expect the mind's composure will make hasty advance. His love was too strong, and the dear object of it too deserving."

To recover some semblance of peace of mind, Babbage soon embarked on a tour of Europe. Though he wished to travel alone, his mother insisted that he be accompanied. With no desire to be served by a valet, Charles chose one of his mechanics, Richard Wright, to travel with him as a colleague. The two men crossed the channel near the end of 1827. Before they left, Babbage instructed his banker to make £1000 available to John Herschel, who would superintend work on the Difference Engine while he was away.

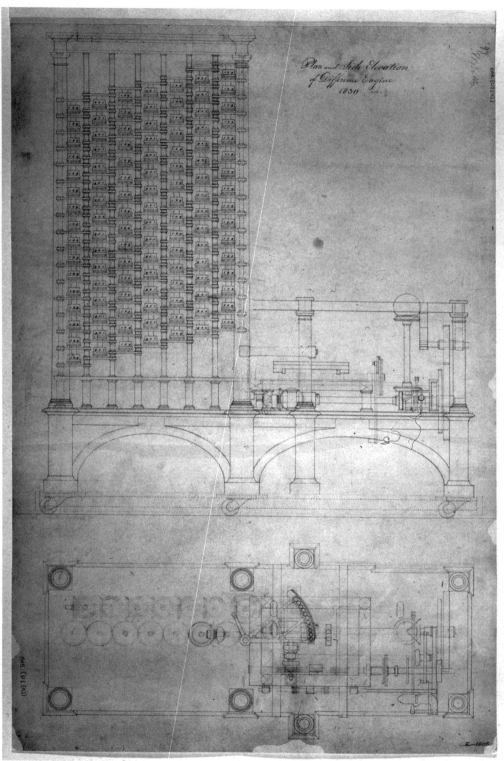

The plan and side elevation of Babbage's Difference Engine No. 1. The physical engine would have measured eight feet high, seven feet wide, and three feet deep.

Inventing the Difference Engine

When Charles Babbage and John Herschel visited Paris in 1819, they inspected a great mathematical work. In the 1790s, Baron Gaspard de Prony had supervised the production of 17 volumes of tables of logarithms and of the trigonometric functions of angles. Though they were never published, the manuscripts were frequently consulted by other table makers. So great a labor could not have been achieved by ordinary methods of calculation. The two Englishmen were surprised to learn that de Prony had devised his unique method after a chance reading in Adam Smith's *Wealth of Nations*. This early book on the principles of industrial economy was published in London in 1776, though Smith was a professor at the University of Glasgow in Scotland. The chapter that impressed de Prony described the division of labor whereby manufacturing processes could be broken into small steps, each performed repetitively by specialized workers.

Baron de Prony applied the division of labor to the production of his mathematical tables. First, a few expert mathematicians decided on the most appropriate formulas to use for the calculations. Second, about eight calculators who knew algebra used the formulas to make detailed calculations of values for the table at regular intervals. A third group calculated all the other values by the method of differences, using only simple addition or subtraction, as instructed by the second group of calculators. Babbage described the work of the third group in his open letter to Humphrey Davy in 1822:

> The third section, on whom the most laborious part of the operations devolved, consisted of from 60 to 80 persons, few of them possessing a knowledge of more than the first rules of arithmetic: these received from the second class certain numbers and differences, with which, by additions and subtractions in a prescribed order, they completed the whole of the tables above mentioned.

A simple example will demonstrate the technique. Suppose you want to construct a table of the squares of integers up to 1000 or more. You consider the task a bore, so you induce a couple of grade schoolers to do the job for you. The only arithmetic they know is addition, but they are good at it. You tell them to add a certain number to another one, add again to the result, and repeat this over and over again. You had better find a good treat to reward them for their labors.

Both Anne and Bob start with the number 1. From then on, Anne will add 2 again and again, passing the results to Bob. Bob, in his turn, will add in the number Anne gives him each time, over and over. The process is shown in the table on the following page.

The numbers in the last column are the squares of the numbers in the first column. All Anne and Bob needed was very simple addition.

The formulas for logarithms and other functions are much more complicated than this. In particular, instead of only two calculators like Anne and Bob in sequence, many more would be needed. That is the kind of work the eight calculators did for Baron de Prony.

Step Number	Anne's Task	Anne's Result	Bob's Task	Bob's Result
1	1	1	1	1
2	1 + 2	3	1 + 3	4
3	3 + 2	5	4 + 5	9
4	5 + 2	7	9 + 7	16
5	7 + 2	9	16 + 9	25
6	9 + 2	11	25 + 11	36
7	11 + 2	13	36 + 13	49
8	13 + 2	15	49 + 15	64
9	15 + 2	17	64 + 17	81
10	17 + 2	19	81 + 19	100
11	19 + 2	21	100 + 21	121
12	21 + 2	23	121 + 23	144
13	23 + 2	25	144 + 25	169
etc.	etc.	etc.	etc.	etc.

Charles Babbage's great idea in 1821 was that the work of the third section could be performed by a machine. All he had to do was to figure out a mechanism that could add constant differences to specified starting values. And that is why he called his machine a *Difference Engine*.

Babbage was convinced that the machine was theoretically possible, though he had no design details. He thought out the basic organization, and began to experiment with mechanisms. His early designs and working models were all hand operated, but the idea of calculation being driven by a steam engine was so appealing that he called his invention the Difference Engine. Developing the full design and constructing it were to be Babbage's main preoccupation for the next decade.

Babbage knew that, for roughly two centuries, famous and ingenious people had worked at constructing calculating machines, some of which actually worked, more or less. So the idea of calculating tables by machine was not very extraordinary. But these hand-operated machines were too slow for the work Babbage envisioned. No adding machine was commercially successful until much later in the 1800s. Since the Difference Engine was never successfully completed, you might conclude that Babbage was an impractical dreamer, especially because he had no prior experience in designing and building complex machinery. You might also conclude that he was foolish to spend so much time and money on his fantastic dream.

However, that is the wrong way to look at the matter. Babbage was wealthy enough not to need financial gain from his work. And he did not know whether his engine would be successful until he built it. While he might hope to contribute to the progress of science and of England, his main drive came from within. His reward came from the intellectual act of invention itself. He could not invent a calculating engine without designing gears, control mechanisms, and power drives. It was not important whether the machine tools of the age could actually produce these parts with sufficiently high quality and low cost to build a working engine.

Babbage created abstract designs, machines existing on paper and in his own mind, rather than in brass and steel. By observing mechanisms closely, and by thinking deeply about them, Charles Babbage made himself into one of the best

text continues on page 40

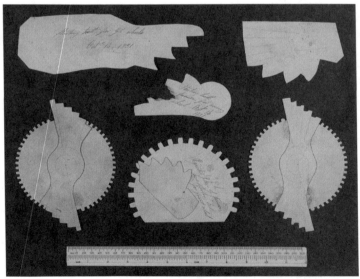

Babbage used cardboard cutouts of various components while developing his designs. Many of the annotations are in Babbage's handwriting and give clues about the contribution made by his engineer, Joseph Clement, to the design process.

DIFFERENCES IN SEQUENCES OF NUMBERS

T o make Babbage's Difference Engine work, the operator has to specify the initial differences to be entered into the machine. For automatic operation, the difference applied to the starting wheel has to be a constant. As you can see in the following tables, in a sequence of the squares of integers, the *second* difference is constant at 2; in a sequence of cubes the *third* difference (6) is constant. Note as a final check, that the next difference after the constant one is zero.

Squares

Sequence	First Difference	Second Difference	Third Difference
1			
4	3		
9	5	2	
16	7	2	0
25	9	2	0
36	11	2	0
49	13	2	0

Cubes

Sequence	First Difference	Second Difference	Third Difference	Fourth Difference
1				
8	7			
27	19	12		
64	37	18	6	
125	61	24	6	0
216	91	30	6	0
343	127	36	6	0

engineering consultants of his time. His initial lack of experience in practical engineering was actually an asset to his basic task. Because he had an unsurpassed genius for abstract mechanical design, his calculating engines were the most complex machines invented before 1900. His ability to do abstract design with little regard to petty problems of implementation allowed him to see the long-term consequences and implications of his ideas more profoundly than perhaps any other figure in the history of technology and engineering.

In these ways, Babbage's unplanned foray into the world of calculating machines was not such a departure from his earlier work in theoretical mathematics and related fields. Lacking a suitable scientific position with any recognition or financial rewards, he had nevertheless created a novel scientific problem, one suited to his particular intellectual style and uniquely his own. No one appointed him computer genius for the next century, and even he did not realize that it would happen. But the vision of automatic computing, as we know it today, was to dominate the rest of his creative life.

Babbage was bold in thinking he could automate a process such as calculating a table of logarithms. He was audacious in supposing he could design a machine to compute practically any mathematical function: not just logarithms, but sines, tangents, square roots, and tables to determine the positions of the moons of Jupiter, or to safely navigate the high seas. Babbage very quickly determined that he would do this by designing the machine around the method of differences.

The basic design of the computing section of the Difference Engine changed very little from its conception. Several vertical columns were arranged across the front of the machine, each holding several rotating horizontal wheels divided into 10 parts, numbered with the digits 0 to 9. Each column corresponded to one complete number, with the most significant digit at the top, and the least significant digit at the bottom. The rightmost column (or axis) contained the table number, the next column the 1st difference, and so leftward through the other orders of difference. Additional mechanisms allowed the digit on any given wheel to be added to the corresponding wheel on the axis to its right, and for "carries" to be propagated up or down a given column when individual wheels passed from 9 to 0 (or the other way). Thus, the basic ability was adding (or subtracting) individual digits. The machine was operated by pulling back and forth on a handle on top, which connected to the internal gears. We can see how this translates to more complex functions by considering a simple case, where we want to compute a table of squares on a Difference Engine with 3 columns (table, 1st difference, and 2nd difference) and where each column has only 3 figure wheels. We will calculate values of a function of a variable. The variable will take the values 0, 1, 2, 3 . . . and so on, and it is represented by the letter x. Our rather simple function is expressed mathematically like this:

$$f(x) = x^2$$

Step 1

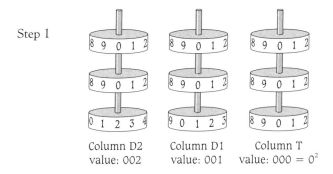

Column D2   Column D1   Column T
value: 002   value: 001   value: 000 = 0^2

The initial set-up should be that shown as Step 1. Here, the value 2 is set in the lowest wheel of the D2 column (which will not change), the value 1 (the initial 1st difference) onto the lowest wheel in D1, and the value 0 (which is the square of zero) into the T (table) column. Here, $x = 0$ and $f(x) = 0$.

Step 2

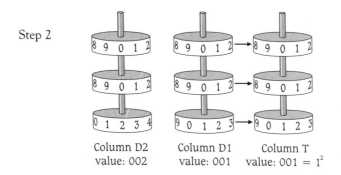

Column D2   Column D1   Column T
value: 002   value: 001   value: $001 = 1^2$

Step 2: Now we are ready to compute by pulling the driving handle back and forth. In the transition to Step 2, each figure wheel in D1 is added to the corresponding wheel in T. In this case, we just add 1 to the lowest wheel, giving what is called Step 2. In more complicated cases, we would have to carry digits upward in Column T, whenever one of its wheels passed from 9 to 0.

Step 3: Now we have a new table value, and can update the D1 value, as shown in step 3, by adding the values in D2 to those in D1.

Step 3

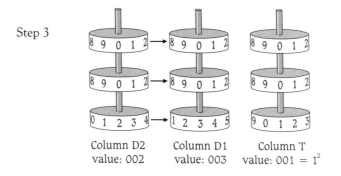

Column D2   Column D1   Column T
value: 002   value: 003   value: $001 = 1^2$

The table values are unchanged in this step. We then add the D1 wheels to the T wheels, leading to step 4, with the value of 2 squared in the table column.

Step 50: Proceeding like this, two steps for each new square (and each new value of x), we would soon reach step 50, with the table column representing the result where $x =$

25 and $f(x) = 625$, and we could continue until we ran out of wheels to hold the results. Because the basic operations are addition and subtraction, it is not much harder to build than a mechanical adding machine. It is, of course, more difficult than these drawings suggest, since they ignore carrying numbers upward, the elaborate provisions for automatic printing of results, and other important details. It is also true that calculating a table of squares by hand is not very difficult, and thus not worth great effort to mechanize.

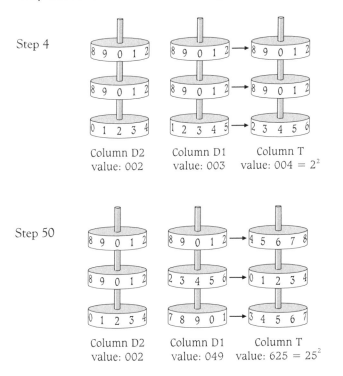

Step 4

Column D2   Column D1   Column T
value: 002   value: 003   value: 004 $= 2^2$

Step 50

Column D2   Column D1   Column T
value: 002   value: 049   value: 625 $= 25^2$

For squares, the engine will need three sets of wheels; for cubes, four. Babbage would need even more sets of wheels in his machine to calculate sequences that might take until the fourth or fifth difference to find a constant value. In fact, Babbage used his mechanical ingenuity to reduce the numbers of sets of wheels to the bare minimum. He described a part of the process late in 1822:

text continues on page 47

The first mechanical calculator we know of was made by Wilhelm Schickard. He was a Professor of Hebrew, Oriental languages, mathematics, astronomy, and geography in the German town of Tübingen, and also a Protestant minister (no narrow specialist, he!). He was also an associate of the great German astronomer Johann Kepler, and we know that he and Kepler had discussed logarithms as early as 1617. Schickard continued working on calculating methods, and in September 1623, he wrote Kepler, saying:

> I have constructed a machine consisting of eleven complete and six partial sprocket wheels that can calculate. You would burst out laughing if you were present to see how it carries by itself from one column of tens to the next, or borrows from them during subtraction.

Kepler clearly was interested because, by the following spring, Schickard was having a second copy of the machine built to send to Kepler. Unfortunately, this version was destroyed when the house in which it was being built burned down. The existence of Schickard's machine was long forgotten, and its details seemingly lost. Then, quite recently, some scholars studying old books in the archives of the Russian Academy of Sciences in St. Petersburg found one of Kepler's own texts with a piece of paper used as a bookmark. It turned out that this paper included Schickard's original drawings of the machine.

The machine itself was quite limited. It allowed addition or subtraction of numbers with up to six digits. Digits were entered one at a time by using a stylus to rotate input wheels through an appropriate fraction of a circle. A simple mechanism automatically carried digits to the left when appropriate in addition, or borrowed from left to right during subtraction. The larger upper section of the machine had parts that would display a multiplication table for a particular number, so that partial products could be added by hand at the bottom.

The carry mechanism worked in a way that would have made it quite difficult to extend the machine beyond six digits. In this form, it would not have been very useful in practical computation. But as a demonstration of the concept of mechanical computation, it was quite elegant.

The first mechanical calculators to be widely known were built by the great French natural philosopher Blaise Pascal, starting in 1642 when he was 19 years old. Although Pascal probably knew nothing of Schickard, his basic mechanism was quite similar, allowing addition or subtraction of multi-digit numbers by rotating input wheels with a stylus.

Pascal had a more complex carry mechanism, one which allowed a machine with many more digits than Schickard's. However, its design did not allow the wheels to be turned backward, so Pascal had a more awkward approach to subtraction than Schickard. Pascal also did not address multiplication at all.

Pascal had some hopes of establishing a profitable business selling such machines, and he experimented at some length with construction methods and materials. Several were built during his lifetime, but the machines were neither very reliable nor very rapid in use. Even though the venture was not a commercial success, it did bring the idea of mechanical calculation to wide attention, and Pascal's efforts were frequently imitated.

Blaise Pascal conceived of his mechanical calculator in 1642 when he was 19. He produced about 50 machines in his lifetime, all based on his early ideas. It is questionable whether any of his calculators were completely reliable.

The most interesting successor to Pascal was the great German philosopher and mathematician Gottfried Wilhelm Leibniz. We know that he became interested in calculating machines after hearing of Pascal's, but it is not clear if Leibniz knew its details. The machine he finally designed (constructed by a Paris clock maker in 1674) was, in any case, completely different, and far more useful. It went far beyond simple addition, with almost fully automated multiplication, using a device called the stepped drum. It was also far more reliable in construction and operation than the Pascal machines. Its design was not surpassed until the end of the 19th century.

It is not fully clear how influential the Leibniz design was, for the physical machine was lost. Sometime late in the 1670s the machine was stored in an attic of a building of the University of Göttingen, where it was completely forgotten (Leibniz evidently having lost interest in it for some reason). It remained there, unknown, until 1879, when a work crew happened across it in a corner while attempting to fix a leak in the roof.

Unlike the Schickard machine, the existence and general capabilities of the Leibniz machine were known through publications, but without much mechanical detail. Some knowledge of the mechanical principles endured—most advanced calculator designs for the next two centuries used Leibniz's stepped-drum mechanism.

Several other interesting prototypes were built in the 1700s, but none really advanced on the functionality of Leibniz's machine. The first commercially successful machine was the Arithmometer, originally designed in 1820 by the Frenchman Charles Thomas de Colmar. The machine was quite similar to Leibniz's. Although a few were available for sale in the 1820s, the machine was very slow to catch on, and was not really successful until after it received wide and favorable notice at the industrial exposition in Paris in 1867.

The first reliable and commercially successful calculator, the de Colmar Arithmometer, was introduced around 1820 and remained in production until around the start of World War I.

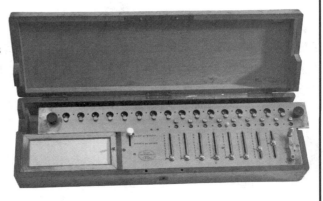

text continued from page 43

According to the original plan, an engine for computing tables—whose second differences are constant and having six figures in each number, and four and two respectively in its first and second differences—would have required 96 wheels and 24 axes. In the reduced engine, 18 wheels and 3 axes became their substitutes. In that part of the engine by which the numbers were to be stamped, a still greater reduction had been effected: 10 dies fulfilled the office of 120.

Within a few weeks of conceiving his idea, Babbage had worked out the main principles of the calculating section of the machine, and he began to think about how to print the results. He was convinced that automatic printing would reduce errors that might occur in copying results or setting them in type.

In May 1822, Babbage put together a working model of a section of the calculating mechanism, including two orders of difference, but no print mechanism. He successfully calculated the first thirty values arising from the formula $x^2 + x + 41$, which was a favorite example of his because it generates a lot of prime numbers. The machine produced correct results at the rate of 33 digits per minute.

In his open letter to Sir Humphrey Davy, Babbage described

a few trials which have been made by some scientific gentlemen to whom [the engine] has been shown, in order to determine the rapidity with which it calculates. The compound table is presented to the eye at two opposite sides of the machine; and a friend having undertaken to write down the numbers as they appeared, it proceeded to make a table from the formula. In the earlier numbers, my friend, in wiriting quickly, rather more than kept pace with the engine; but as soon as four figures were required, the machine was at least equal in speed to the writer.

This was the stage of development that Babbage had reached when he announced his machine to the public in mid-1822.

British engineer Mark Brunel's tunnel under the Thames River linked north and south London. Construction took 18 years and cost £300,000. Babbage visited the tunnel just before leaving on his tour of continental Europe.

Reform Is in the Air

Just before Babbage and Wright left England at the end of 1827, Charles took his eldest son, Herschel, age 12, to visit a great engineering work. Since 1825, Marc Brunel had been building a tunnel under the Thames River in London's east end. This was a massive undertaking—when completed, the tunnel was 1500 feet long, 37 feet wide, and 23 feet high. Because of many technical and financial difficulties, the tunnel was not completed until 1843, at a cost approaching £300,000. Charles may well have wondered how the government could afford so much for a tunnel and be so stingy over his Difference Engine.

The Babbages were shown the tunnel site by Marc's son, Isambard Kingdom Brunel, who was overseeing the work though he was only 21. Ten years later, Herschel was employed by Isambard on the construction of the Great Western Railway. For now, Charles could use the experience to impress his friends on the continent. He bought a dozen copies of a description of the tunnel project. As he wrote,

> Six of the copies were in French and the other six in the German language. I frequently lent a copy, and upon some occasions I gave one away; but if I had had twice that

number I should have found that I might have distributed them with advantage as acknowledgements of the many attentions I received.

Charles Babbage had learned that tourists are treated better when they have gifts to exchange for the hospitality shown them by their hosts in foreign lands.

Parting from his broken family, Charles and his friend went first to Holland. They traveled at a leisurely pace, visiting scientists and artisans along the way. They passed through Belgium into western Germany, and then turned southward. On the route from Frankfurt to Munich they shared a ride with a young man whose father was the coachmaker of the Tsar of Russia. This son was searching for information on the best techniques for making coaches and carriages. During the trip, Babbage learned from him every part of the structure of a carriage. He made careful notes of the details, so that, by the time they arrived in Munich, Babbage knew enough to be able to design his own carriage. The young Russian enjoyed Charles's company so much that he invited him to return to Russia with him. However, Charles declined, wanting to get to Italy as soon as possible.

Babbage and Wright crossed through the Brenner Pass in the Alps into Italy and spent some days visiting factories in Venice. Charles was gratified to learn from one metal worker there that files metal manufactured in Lancashire were the best obtainable. Then, on to Bologna, where they spent several weeks in discussions at the university and with various craft experts. After that, in Florence, Babbage met the Grand Duke of Tuscany, with whom he became very friendly. He asked Babbage if there was anything his government could do to advance Italian science. Charles recommended holding regular scientific congresses, so that scientists could consult with one another about their work. The Grand Duke was impressed but took a dozen years to make it happen. When he did, Babbage was invited to attend.

The two Englishmen went next to Rome. There, one day in the spring of 1828, Charles was surprised to see the following note in a local newspaper: "Cambridge, England. Yesterday the bells of St. Mary rang on the election of Mr. Babbage as Lucasian Professor of Mathematics." This university chair, once held by Isaac Newton, was a great honor, though it carried an annual salary of less than £100. Soon, two English friends came by to congratulate Babbage. He told them that he had just drafted a reply refusing the appointment. He did not think it was worth the distraction from his beloved Difference Engine. However, his visitors pointed out that the masters of the Cambridge colleges who had chosen him, as well as his own friends who had influenced them, would be offended by being rejected. To this he had no satisfactory reply. He accepted the post and held it for ten years. However, he did not live in Cambridge and seldom lectured there.

Babbage had intended to extend his trip into Asia. But the Greek war of independence from the Ottoman Empire made travel risky between southern Europe and Asia. Instead, he proceeded to Naples. There, Babbage showed his scientific versatility by turning to investigations in geology. He found a guide who would take him to the summit of the volcano Vesuvius, which was moderately active. He persuaded a member of the team to accompany him down to the bottom of the crater, 600 feet below the rim. With instruments he carried along, Babbage measured temperatures and atmospheric pressures along a grid on the crater floor. One section was bubbling gently, and Babbage crept close enough to look down into the sea of lava. By the time he returned to Naples he found that his thick boots were falling apart from the heat they had endured.

While he was in Naples, its government appointed a commission to report on the extent and usefulness of hot springs on the island of Ischia off the coast. That Babbage was made a member of this team shows the high recognition he received on foreign soil—so much more, he felt, than at home in England. Babbage considered the possibility that the hot springs might be exploited as a source of power.

From Naples, Babbage and Wright traveled back north through Italy. After spending a couple months in Florence, they continued on through Venice and on to Vienna. There, as Charles later wrote, he bought a carriage.

> I had built for me at Vienna, from my own design, a strong light four-wheeled calèche in which I could sleep at full length. Amongst its conveniences were a lamp by which I occasionally boiled an egg or cooked my breakfast; a large shallow drawer in which might be placed, without folding, plans, drawings, and dress-coats; small pockets for the various kinds of money, a larger one for traveling books and telescopes, and many other conveniences. It cost somewhat about £60. After carrying me during six months, at the expense of only five francs for repair, I sold it at the Hague [in Holland] for £30.

Babbage made no mention of the arrangements for the horses needed to pull his carriage. You can see the kind of style a well-to-do gentleman could travel in back then.

Babbage and Wright then set out across Germany to Berlin. There, Babbage was eager to meet Europe's leading scientist of the century, Alexander von Humboldt. An explorer on three continents, Humboldt was a tireless observer and collector in geology and biology. He also made great contributions to meteorology and the study of Earth's magnetism. Moreover, the king of Prussia sent him on frequent diplomatic missions.

When Babbage arrived in Berlin, Humboldt was planning the seventh annual congress of German scientists. He put Babbage on the committee to investigate the restaurants they should use for feeding the delegates. According to Humboldt, Englishmen always appreciate a good dinner. The congress opened in mid-September 1828, with almost 400 delegates from throughout central Europe. The opening ceremonies were attended by an additional 800 local digni-

Baron von Humboldt, a German naturalist, founded the studies of physical geography and meteorology, and traveled extensively in South America and central Asia.

taries. The scientists included Hans Christian Oersted, the Danish discoverer of electromagnetism, and Karl Friedrich Gauss, the foremost mathematician and physicist of his time.

All this evidence of the high status accorded to science in Europe impressed Babbage immensely. He resolved to use his experience to promote the cause of science more vigorously when he returned to England. And it was not just the cause of science. Babbage fervently believed that science could be applied to the improvement of social conditions. Both Britain and Europe were still ruled largely for the benefit of the wealthy landowning class. Babbage's friends in Europe were active in changing that, to give ordinary working people a larger role in their own government.

Their views fitted well with his own liberal attitudes regarding conditions back home.

Returning to London near the end of 1828, Babbage threw himself into an astonishingly wide range of activities. He took the chair of Lucasian professor of mathematics at Cambridge. From 1829 to 1834, he engaged in electoral politics, promoting candidates and even standing for election himself. He also began a campaign for the reform of the Royal Society, which failed, leading him to promote the formation of a new scientific organization. In addition, he looked after the affairs of his family, continued with the Difference Engine, and managed to write a 400-page book on the economy of manufacturing. Consider each of these in turn.

Babbage had few duties as Lucasian professor at Cambridge. Occasionally, he was required to act as examiner in special mathematical examinations. He gave no lectures and had no students. Even so, by 1839 he felt that designing calculating machines demanded so much time that he resigned the post. Yet he was always grateful for the appointment, which he called "the only honour I ever received in my own country."

In those days, the two-house parliament of England was in the hands of the landowning gentry and aristocracy. Dukes and bishops formed the House of Lords. And although elections were held to fill the House of Commons, many of those places were also controlled by the dukes and bishops. For 50 years, as the Industrial Revolution gathered steam, many peasants had moved to factory towns such as Birmingham and Manchester. By 1830, these towns had become large cities, but they had no representation in parliament. Some of the older country towns had shrunk almost out of existence, but still sent members to parliament—they came to be called "rotten boroughs."

The new industrialists—factory owners and investors—desired representation in parliament. They wanted laws that favored their economic needs rather than those of the landowners. For example, a law in the 1820s put high tariffs on imported grain to keep the prices high and less competitive with those of local producers. So, in the elections of 1829 and 1831, Charles Babbage became chairman of a campaign committee to elect a reforming member of parliament, representing Cambridge University. Charles worked energetically and effectively—his man won the first election.

In the new parliament, a Reform Bill was introduced to reassign 143 seats from rotten boroughs to the new centers of population in the English midlands. It aroused such strong opposition among conservative members and the House of Lords that parliament was dissolved, and a new election was called in 1831. For that election, Babbage not only worked for his Cambridge candidate—who lost—but also for his brother-in-law, Wolryche Whitmore in Shropshire. He won. While in Shropshire, Babbage actively supported the election of other liberals in the area. Although they lost, the liberals did achieve a majority in the new House of Commons. The Reform Bill passed in the Commons but was rejected by the House of Lords. Riots against that result broke out in major population centers. Frightened, King William IV persuaded the Lords to change their minds. The Reform Bill substantially enlarged the franchise, though Britain was still far from having universal suffrage. Poor working men could not yet vote, and women were not enfranchised until well after 1900.

The passage of the Reform Bill led to a new election based on the redistribution of seats in the House of Commons. Babbage's friends persuaded him to seek a seat. He agreed to stand for the new constituency of Finsbury in north London. The *Mechanics Magazine* published an article strongly supporting Babbage's bid for election:

> We look for a great deal of good to science, as well as to every other important interest of the country, from the return to Parliament of a gentleman of Mr. Babbage's emi-

nence in the scientific world, tried independence of spirit and very searching and business-like habits; and therefore we take the liberty to say to every elector of Finsbury who is a reader of this journal and a friend to the objects it has especially in view—Go and vote for Mr. Babbage. If you are an inventor, whom the iniquitous and oppressive tax on patents shuts out from the field of fair competition, and are desirous of seeing that tax removed—Go and vote for Mr. Babbage. If you are a manufacturer, harassed and obstructed in your operations by fiscal regulation—Go and vote for Mr. Babbage. If you are a mechanic, depending for your daily bread on a constant and steady demand for the products of your skill, and are as alive as you ought to be to the influence of free trade on your fortunes—Go and vote for Mr. Babbage.

Mr. Babbage, with 2,311 votes, lost the election by 537. Two years later, at a by-election, he was again defeated. That finished his career in electoral politics.

The politics of science would keep him busy enough. Babbage became active in attempts to reform the Royal Society. In 1829, he published a report about the scientific congress he had attended in Berlin. He held it up as an event worthy of emulation in England. A year later, he fired a broadside directly at the Royal Society in a book, *Reflections on the Decline of Science in England, and on Some of its Causes.* One of his major complaints was the nature of the membership of the Royal Society. Right from its founding in 1662, the society admitted to its fellowship many men—no women were admitted until after 1920—with no scientific training. In 1830, nonscientists numbered 450 of the total 660 the Royal Society fellows. Of the 210 with scientific training, half were medical practitioners, many of whom engaged in no research. Babbage felt that the real scientists were swamped in a sea of sociable amateurs.

What was worse, Babbage felt that the nonscientists exerted far too much influence on the society's affairs. Presidents Joseph Banks and Humphrey Davy, although

Sir Joseph Banks, an English naturalist, traveled around the world as a botanist with Captain James Cook. He was president of the Royal Society from 1778 to 1820.

themselves scientists, had used that influence to dominate the society with their own personal views. They appointed the members of the governing council and dispensed benefits to their friends. That was no way to promote vigorous scientific activity in England. Humphrey Davy died in 1829. In the election to replace him, Babbage and his friends put forward John Herschel. To oppose him, the Davy party proposed the Duke of Sussex, a younger brother of King George IV and King William IV. The duke received 119 votes from the nonscientific fellows. Herschel lost with 111 votes from the scientific fellows.

Significant reform in the Royal Society would not come for another 20 years. Babbage was impatient. Having no faith in the London establishment, he consulted with scientific friends across England and Scotland. Together, they formed the British Association for the Advancement of Science—often called the BA. Their model was the scientific congress of Germany. The association's major objective was to hold annual meetings of scientists at various centers throughout Britain. The first meeting was held in York in northern England during the summer of 1831, with 350 in attendance. Charles Babbage became one of three permanent trustees of the association.

As the British Association prospered, its meetings were divided into sections, each devoted to a particular branch of science. At Cambridge in 1833, Babbage organized a statistical section, of which he became chairman. A year later, he also helped to found the Statistical Society of London,

Three months (April–June 1844) in the social diary of the Babbages indicate a full schedule, including several meetings with the Duke of Somerset, a close friend, as well as portrait sittings with the artist Samuel Laurence.

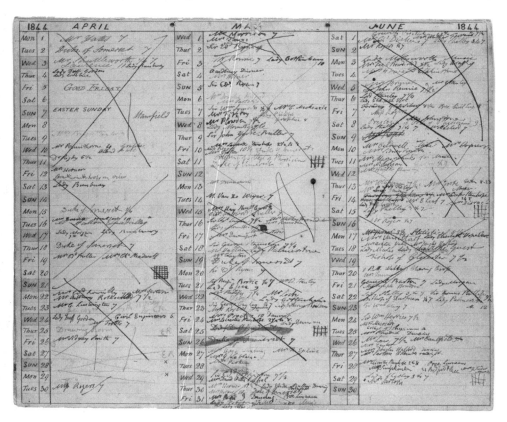

independent of the BA, to promote the gathering and analysis of information about the progress of the British economy.

Soon after Charles had returned to England in 1828, he decided he could afford larger premises. He leased a second house several blocks from where his mother was tending his children. He had ample space for workshops, but also had a fine suite of rooms for entertaining guests. He began to hold regular Saturday evening parties, initially in order to introduce his teenage children, Herschel and Georgiana, into society.

Before long, the Babbage soirées formed an important part of the London social scene. Often, the guest list exceeded 200. They came from all parts of polite society: lawyers and judges, doctors and surgeons, deacons and bishops, and scholars and artists by the score. There were aristocrats like the Duke of Wellington, hero of Waterloo, and the Marquis of Lansdowne, a reforming minister in Liberal cabinets. From the arts and letters came Shakespearian actor William Macready, historians Thomas Macauley and Henry Milman, the novelist Charles Dickens, and the celebrated wit Sydney Smith. The scientists included telegraph inventor Charles Wheatstone, geologists Charles Lyell and William Fitton, and the young biologist and world traveler, Charles Darwin. Photographic inventor William Fox-Talbot came with his friend John Herschel. Visitors from abroad were also welcomed: the German composer Felix Mendelssohn; Camillo Cavour, the Italian statesman who was later active in the unification of his country; Alexis de Tocqueville, the French author of *Democracy in America*; and from America, the physicist Joseph Henry.

Wives came with their husbands, and some women were welcomed for their own special qualities. The heiress Angela Burdett-Coutts was herself a splendid host at garden parties that Babbage attended with pleasure. Angela took up the study of astronomy, and Charles often accompanied her to the lessons. During the Crimean War, Angela purchased equipment to send to nurse Florence Nightingale. Mary Somerville, wife of a physician, enlivened the parties with her deep understanding of science. She wrote several outstanding books on science for the general public.

All this time, Charles continued to direct the production of his beloved Difference Engine. As it turned out, things were not going so well. By 1828, Charles had spent more than £6000 on the construction, and the government had only reimbursed him for £1500. After a supportive report from Charles's friends in the Royal Society, the government agreed to make up the difference. But the work went slowly. The engineer, Joseph Clement, was refusing to continue unless he was paid promptly; and government vouchers had to wend their way through a complex bureaucracy.

The whole project was taking much longer than anyone had anticipated. Being the first one of its kind, it had significant growing pains. While the fabrication of basic parts proceeded, shop patterns had to be drawn for others. The full set of plans was not completed until 1830. By then, Clement's workers had produced many thousands of parts, but had done little assembly. Soon, Babbage and the government decided that the plans and assembly should be moved out of Clement's shop. On Babbage's new property, they built a two-storey fireproof workshop 60 feet long by 25 feet wide. A second building would house the Difference Engine. Babbage's intention was to move Clement's whole operation to these new quarters. However, Clement resisted. With the funds Babbage had supplied him, he had greatly expanded his own workshop. He now had many machine tools and a number of employees. He used them to do other work besides that contracted by Babbage. And by the trade practices of the time, he insisted that the machinery belonged to him, not to either Babbage or the government.

During 1832, Clement's workers completed the assembly of as much of the engine as they had parts for. Even

This portion of Difference Engine No. 1, assembled by Joseph Clement in 1832, is the first known automatic calculator. The portion shown has nearly 2,000 individual parts, and is one of the finest examples of precision engineering of the time.

though the calculating section was largely complete, but the printing section was not. From this time on, no further work was done. Clement would not move his machinery to Babbage's shop, and only in 1834 was the engine itself transferred. By then, the government had expended £17,000, and Babbage had spent several thousand pounds more. The government was unwilling to proceed further, given the need to reorganize the whole project after Clement and Babbage had parted company.

For some years, Babbage displayed the working section of his Difference Engine in one of his drawing rooms. In an adjoining room, he had the dancing figurine that he had recently bought and refurbished. During one jovial Saturday evening party, Babbage watched a large crowd of friends admiring the figurine's graceful movements. Next door, two foreign visitors, an American and a Dutchman, were earnestly discussing the workings of the Difference Engine. Frowning, Babbage remarked to a friend, "Here [watching the toy] you see England; there, two foreigners." He clearly felt the pangs of the lack of recognition of his achievements in his own country.

Babbage turned every experience to advantage. After all his visits to workshops and factories both in England and on the continent, he sought to draw general principles from them. In 1832, Babbage compiled these principles into the more than 30 chapters of his book *On the Economy of Machinery and Manufactures*. Within three years, there were four editions in England, one in America, and translations into German, French, Italian, Spanish, Swedish, and Russian—a genuine best-seller.

In his *Economy*, Babbage analyzed industrial production in all its aspects: from obtaining and transporting raw materials, through the proper location and arrangements of machinery and workers, to the distribution and cost accounting of the finished product. He illuminated each step along the way with principles and examples. He also paid attention to the relations between labor and management. His objective was to recommend economies and efficiencies. With the Industrial Revolution only about 50 years old, Babbage provided an important and useful blueprint for its future. Unfortunately, in many cases, those with vested interests did not care to introduce procedures that would benefit workers and the buying public. In other cases, Babbage made the kinds of recommendations that would become the stock in trade of efficiency experts in our own century.

Babbage was fully aware of the dislocation that new manufacturing techniques brought to working families. As power-weaving machinery replaced hand-loom weavers, strong male weavers were replaced by women and children as machine tenders. So Babbage made the following suggestions:

Increased intelligence amongst the working classes may enable them to foresee some of those improvements which are likely for a time to affect the value of their labor; and the assistance of savings banks and friendly societies (the advantages of which can never be too frequently, or too strongly, pressed upon their attention) may be of some avail in remedying the evil; but it may be useful also to suggest to them that a diversity of employments amongst the members of one family will tend, in some measure, to mitigate the privations which arise from fluctuation in the value of labor.

Keeping something in the bank for a rainy day and developing a variety of skills are the kinds of advice we can still use in today's changing labor market.

An example of Babbage's idea of efficiency occurs in his analysis of mail transport between London and Bristol. To send a 100-pound sack of letters by horse-drawn coach required the effort to pull a two-ton vehicle. So Babbage suggested that a system of elevated wires be erected between London and Bristol. With the letters in a light metal container on wheels running on the wires, the effort of transporting the mail would be greatly reduced. Such a system was never used over long distances, but was used in department stores in the 1920s. Babbage would certainly have been overjoyed by e-mail, by which electronic symbols are transported all over the world in the blink of an eye.

In the midst of this full bustle of activity during the 1830s, personal tragedy again struck Charles Babbage. In 1834, his beloved daughter Georgiana became ill and died. She was only 17 years old. To deal with his grief, he threw himself more deeply into his work. His son Herschel moved into his home for a while, and his two younger sons left their boarding school to live with their grandmother in the other house.

An important innovation at this time was the development of the English railway system. Beginning with relatively short point-to-point lines, initially intended for the transport of such goods as coal, the railways had more than

6,000 miles of line by 1850. Passenger traffic soon rivaled freight. Babbage and his friends could travel to British Association meetings in steam-driven trains instead of horse-drawn coaches. And soon working-class families could afford a summer holiday at the seaside.

In September 1830, Babbage and his brother-in-law Wolryche Whitmore were among the dignitaries at the opening of the Manchester to Liverpool railway—covering a distance of about 40 miles. For the next ten years, Babbage became progressively more involved in developing the efficiency of rail transport. He wrote a letter of introduction for Isambard Brunel to the projectors of a railway from Birmingham to Bristol. Brunel was appointed engineer of the project, but financing fell through. In 1833, Brunel did get a similar post for the London to Bristol railway, which he named the Great Western Railway. Then, in 1837, to extend the trip all the way to New York, Brunel built the paddle-wheeled steamship, the *Great Western*. To get some idea of engineering progress at the time, consider that, 20 years later, Brunel built the *Great Eastern* of 32,000 tons, more than ten times the tonnage of the *Great Western*.

Always a faithful friend to Brunel, Babbage had an opportunity to do a great service for him a few years later. When Brunel designed the Great Western Railway, he chose 7 feet for the distance between the tracks—what is called the *gauge*. All the previous railways had the standard gauge of 4 feet, 8-1/2 inches. Traditionalists challenged Brunel's choice at meetings of shareholders, where Babbage defended him. In 1838, Charles gave up vacation time to travel on seven railway lines to investigate the extent of their uncomfortable vibrations. He reported that the ride on the Great Western was second-best in quality, noting that it had been traveling at 40 mph, while the others averaged only about 15 mph.

The Great Western directors authorized Babbage to make a more detailed analysis. He equipped a coach with a set of measuring instruments. These were designed by his son Herschel, now an employee of Brunel's. The instruments recorded the speed of the train and its degree of vibration in all directions. Babbage presented his results at the next stockholders' meeting and won a resounding success. He found the results so valuable that he suggested such instruments be a permanent feature on all trains. In the case of accident, they would help to determine its cause. Nowadays, using electronics, trains and aircraft are equipped with "black boxes" like that. However, in the long run, tradition won out. By 1900, despite the superiority of the 7-foot gauge, it had all but disappeared.

By the time work on the Difference Engine had ceased, Charles Babbage's ingenious mind was already working on a vastly improved device. What he called his Analytical Engine would do a lot more than merely generate numbers by adding or subtracting fixed amounts—it would solve equations. In the Difference Engine, whenever a new constant was needed in a set of calculations, it had to be entered by hand. In 1834, Babbage conceived a way to have the differences inserted mechanically. But he wanted a machine that could solve more complicated problems—he wanted a computer.

For more than 20 years, Babbage labored to design the various sections of the Analytical Engine to produce what would have been a mechanical (rather than electronic) computer. What we call the CPU (central processing unit), he called a mill. What we call memory, he called the store. Instead of electrical signals in conductors connecting the sections, Babbage had a variety of gears and levers.

Lacking both support and encouragement from the government, Babbage embarked on designing the Analytical Engine with his own funds. He started his workshops with a forge, some machinery, and an elaborate drafting room. He hired C. G. Jarvis, who had been Clement's draftsman. By now, Babbage was realistic enough to know

text continues on page 68

THE OPERATION OF THE JACQUARD LOOM

This passage is from Luigi Menabrea's paper describing Babbage's Analytical Engine. It was translated by Ada Lovelace and published in London in 1843.

Two species of threads are usually distinguished in woven stuffs; one is the *warp* or longitudinal thread, the other the *woof* or transverse thread, which is conveyed by the instrument called the shuttle, which crosses the warp. When a brocaded stuff is required, it is necessary in turn to prevent certain [warp] threads from crossing the woof, and this according to a succession which is determined by the nature of the design that is to be reproduced. Formerly this process was lengthy and difficult, and it was requisite that the workman, by attending to the design which he was to copy, should himself regulate the movements the threads were to

take. Thence arose the high price of this description of stuffs, especially if threads of various colors entered into the fabric. To simplify this manufacture, Jacquard devised the plan of connecting each group of threads that were to act together, with a distinct lever belonging exclusively to that group. All these levers terminate in rods, which are united together in one bun-

This portrait was woven using a Jacquard loom controlled by punched cards. The cards, strung together in a sequence, determined the weave. A small model of the Jacquard loom appears on the left behind the seated figure.

dle, having usually the form of a parallelopiped with a rectangular base. The rods are cylindrical, and are separated from each other by small intervals. The process of raising the threads is thus resolved into that of moving these various lever arms in the requisite order. To effect this, a rectangular sheet of pasteboard is taken somewhat larger in size than a section of the bundle of lever arms. If this sheet be applied to the base of the bundle, and an advancing motion be then communicated to the pasteboard, this latter will move with it all the rods of the bundle, and consequently the threads that are connected with each of them. But if the pasteboard, instead of being plain, were pierced with holes corresponding to the extremities of the levers which meet it, then, since each of the levers would pass through the pasteboard during its motion, they would all remain in their places. We thus see that it is easy so to determine the position of the holes in the pasteboard, that, at any given moment there shall be a certain number of levers, and consequently of parcels of threads, raised, whilst the rest remain where they were. Supposing this process is successively repeated according to a law indicated by the pattern to be executed, we perceive that this pattern may be reproduced on the stuff. For this purpose we need merely compose a series of cards according to the law required, and arrange them in suitable order one after the other; then, by causing them to pass over a polygonal beam which is so connected as to turn a new face for every stroke of the shuttle, which face shall then be impelled parallelly to itself against the bundle of lever arms, the operation of raising the threads will be regularly performed. Thus we see that brocaded tissues may be manufactured with a precision and rapidity formerly difficult to obtain.

that actually building a complete analytical engine was beyond his capacities and resources. What he could do, and did, was to work tirelessly in research and development—solving the myriad technical problems, making complete engineering drawings, fabricating sets of sample parts, and constructing various components to show details of some of the workings. In addition, he could communicate his ideas widely, on the off chance that someone else might carry the work further.

Babbage made one further technical addition to the Analytical Engine. To input numbers into his engine, he used punched cards—a series of cards with holes in them to represent the numbers. This procedure had been invented in France a hundred years earlier and, around 1800, it was perfected for weaving intricate patterns. The Jacquard loom used punched cards to control the raising and lowering of the warp threads differently on each pass of the shuttle. As Ada Lovelace, a good friend of Charles's, would write later, "the Analytic Engine *weaves algebraic patterns* just as the Jacquard loom weaves flowers and leaves."

Ada Lovelace was born Augusta Ada, only child of the poet Lord Byron. A month after she was born in December 1815, her mother and father separated. She would never know her father, who died in 1824. Lady Byron had training in mathematics, in which she also encouraged Ada. In 1832, Ada met Mary Somerville, who helped advance her study of mathematics. Mary also introduced her to William King, soon to become the Earl of Lovelace. He and Ada were married in 1834 and had three children. In comfortable circumstances, Ada Lovelace spent more time in mathematical and social circles than in raising her children. Her mother and servants looked after them.

Ada Lovelace met Charles Babbage in 1833 and immediately became deeply fascinated by his calculating engines. They became lifelong friends. Ada was only two or three years older than Charles's daughter Georgiana. After her death, Ada and Charles established the daughter–father relationship they both now lacked. They frequently visited each other's homes. In the early 1840s, Ada was to make an important contribution to public knowledge about Babbage's Analytical Engine.

In 1840, Babbage made another journey to the continent. In Lyon, France, he visited a silk-weaving plant that was using Jacquard looms. He watched in great fascination as the loom with 24,000 punched cards automatically generated a very fine portrait of the inventor, Joseph Jacquard. Charles obtained two copies of it. Later, he hung one in his drawing room to amaze his friends. He wrote that this "sheet of woven silk, framed and glazed, looked so perfectly like an engraving that it had been mistaken for such by

This rare daguerreotype shows Ada Lovelace in 1844, about the time of publication of her translation of Luigi Menabrea's memoir on the Analytical Engine. "The Analytical Engine weaves algebraic patterns just as the Jacquard loom weaves flowers and leaves," she said.

two members of the Royal Academy [of painters and illustrators]." This was the ingenuity of the punched-card system of control that Babbage used for his Analytical Engine.

Babbage continued on to Turin to attend the second congress of Italian scientists, which he had urged upon them some years earlier. At the congress, Babbage spent many hours describing his Analytical Engine to Italian mathematicians. Responding to their questions, he found his ideas becoming clarified as he was forced to find explanations that would satisfy others. During these sessions, a young mathematician, Luigi Menabrea, took copious notes. With further assistance from Babbage, Menabrea published a 24-page description of the Analytical Engine (in French) in a Swiss journal in 1842. Later, Menabrea was active in the fight to unify the Italian states, and for two years in the 1860s was prime minister of the new Italian government.

Back in England, Charles Wheatstone suggested to Ada Lovelace that she translate Menabrea's article into English. She agreed, and at Babbage's urging added additional notes. They extended to twice the length of the translation. The article with notes was published in the London journal *Scientific Memoirs* in 1843. In her notes, under Charles's guidance, Ada gave additional explanations and more details of examples to show the power of the Analytical Engine. Several of her remarks have a modern ring that can still be applied to today's computers:

> The Analytical Engine has no pretensions whatever to originate anything. It can do whatever we know how to order it to perform. It can follow analysis; but it has no power of anticipating any analytical relations or truths. . . . [However], in distributing and combining the truths and the formulas of analysis . . . the relations and the nature of many subjects in that science are necessarily thrown into new lights, and more profoundly investigated.
>
> The engine can arrange and combine its numerical quantities exactly as if they were letters or any other general symbols; and in fact, it might bring out its results in algebraic notation, were provisions made accordingly.
>
> Again, it might act upon other things besides number, were objects found whose mutual fundamental relations could be expressed by those of the abstract science of operations . . . Supposing, for instance, that the fundamental relations of pitched sounds in the science of harmony and of musical composition were susceptible of such expression and adaptations, the engine might compose elaborate and scientific pieces of music of any degree of complexity or extent.

Imagine how much Ada and Charles would have loved word processing, spreadsheets, and databases!

During the 1830s, Charles's younger sons attended London's University College for a while. They also spent time in their father's workshop and learned his mechanical notation from the draftsman, Jarvis. Their elder brother, Herschel, married in 1839, in a match that Charles disapproved. Did he have to emulate old Benjamin? Mutual friends helped to smooth things over. When Herschel, with his family and his brother Dugald, went off on a railway project in Italy in 1842, Charles helped them pack up. After other jobs, these two sons went to Australia in 1851 to conduct a geological survey. The third son, Henry, decided to join the Indian army. He took up his post there in 1843. Charles's mother, Betty, was left alone in the old house. She died in 1844 in her mid-eighties.

Charles fell into a routine that lasted most of the rest of his life. He devoted mornings and afternoons to writing or work on the Analytical Engine, and then evenings to dinner, followed by a party, a play, or the opera. His son Henry wrote later that, during the month of February 1842, his father had at least 13 invitations to dinners or parties for every day, including Sundays. And he continued to entertain at home. The Scots chemist Lyon Playfair described a day he spent there:

> Babbage was full of information which he gave in an attractive way. I once went to breakfast with him at 9 o'clock. He explained to me the working of his calculating machine, and afterwards his method of signaling by flashing lights. As I was engaged to lunch at 1 o'clock, I looked at my watch, which indicated the hour of 4. This appeared obviously impossible so I went into the hall to look for the correct time, and to my astonishment that also gave the hour as 4. The philosopher had in fact been so fascinating in his descriptions and conversation that neither he nor I had noticed the lapse of time.

Apparently, Babbage was not tied to his desk every day. However, even explaining complicated matters to friends can be work.

This portion of the mill of the Analytical Engine includes a printing mechanism. Babbage knew that the design of the Analytical Engine would evolve over time.

Inventing the Analytical Engine

Around 1834, Charles Babbage began to design a machine that would overcome a major limitation of his Difference Engine. That machine could calculate a table of numbers for only a single manually entered difference. If the difference needed to be changed to fit a formula, the machine had to be adjusted to take that new value. However, there are many useful formulas in which the difference does change frequently, including those for logarithms and the functions of angles in trigonometry.

Babbage looked for a way to to deal with this problem. He found one method fairly quickly. For calculating a table of the sines of angles, he realized (from trigonometry) that the second difference was a simple function of the value of the sine just calculated. So, all he needed was a way to feed back the value of that sine (multiplied by a constant factor) from the table axis to the second difference axis of the machine. Then, successive values of sines could be calculated without human intervention. Babbage had already understood this much in 1822. In fact, he built extra wheels into the Difference Engine he assembled in 1832. He used them to demonstrate the above principle by allowing a single digit of the table value to be fed back to the second difference column.

Babbage had not attempted to build this more general method into the Difference Engine for a simple reason. A key feature of the method of differences is that it allows any complex function to be tabulated using only addition, which is easy to mechanize. But feeding values back from the table axis would require multiplication as well, which would make the machine much slower and more complex. Babbage chose a design he believed could actually be built.

Babbage called his method of feeding back numbers from one axis to another "the engine eating its own tail."

Babbage kept informal workbooks that he called "scribbling books." These documents of designs and exploratory schemes total nearly 7,000 pages of manuscript.

And he soon began thinking about ways to extend it beyond the single-digit capability in the demonstration section actually assembled in 1832. The following sketch appears in the engineering workbook he referred to with the whimsical title "Great Scribbling Book."

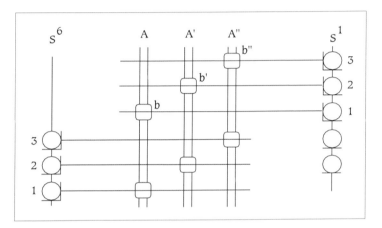

The drawing was labeled "Plan for multiplying any numbers on any s axis and adding them to any other." In this case, he shows the first difference axis, s_1, on the right, and the sixth difference axis, s_6, on the left. In between, Babbage has added three new axes, A, A', and A". Miter gears connect the top three wheels of the s_1 axis to three horizontal rods, which then connect to and turn the racks b, b', and b". These turn the axes A, A', and A", which turn the horizontal rods at the bottom and also the three wheels, labeled 3, 2, and 1, on s_6. A three-digit number has been transferred from s_1 to s_6, multiplying it by 1,000 by shifting it down three levels.

This method would still not have allowed calculation of any genuinely useful tables because an equation such as $\sin(x) = K \sin(x + i)$ requires multiplication of one result by an arbitrary multi-digit constant K, rather than just an integral power of ten. However, this did not keep Babbage from exploring the concept further. His first idea was closely linked to the image of "the engine eating its own tail."

He could eliminate the long horizontal rods in the sketch if he simply arranged the axes of the engine in a circle rather than a straight line. That way, the table axis could be connected quite directly to the lowest difference axis. But this was too confining, for he wanted to be able to interconnect any two axes.

So he combined the two schemes. The original axis of the Difference Engine became several "adding axes," so called because gears allowed the value on one of them to be added to the adjacent axis. The axes labeled A, A', and A" in the sketch he called "multiplying axes" because they allowed numbers to be multiplied by an integral power of ten, by stepping digits up or down between levels. Both adding and multiplying axes were arranged in a large circle, around large central gears with which they could be selectively connected. Thus, any single adding axis could be connected to and drive one set of central gears. The multiplying axes could step the first set of digits up or down to a second set of central gears, which in turn could drive any desired adding axis.

Babbage reached this stage in the fall of 1834. It can be called the final stage of the Difference Engine. Babbage soon realized his plan had potential that completely superseded the Difference Engine. For it to be really useful, multiplication would have to be possible with any combination of digits, rather than just integral powers of ten. A simple method of mechanizing multiplication through adding and shifting had been worked out by Leibniz and Thomas de Colmar. But Babbage's ambition was greater, for he wanted multiplication to be fully automatic and to allow the value stored on any one axis to be multiplied by that on any second axis, with the result stored on any third axis. Babbage's control mechanism began to get complex.

It got even more complex when Babbage extended the machine to do automatic division. This could be accomplished by repeated subtraction and stepping (shifting the divisor by a factor of 10). But it required additional hardware to allow the divisor to be compared at any point with the remainder to see which was larger and determine if it was time for stepping.

The question of how to implement multiplication and division was one that Babbage shared with modern computer designers. So, too, was the problem of how to perform carries. Each time a cycle of addition was performed, the result of adding two digits at any single level might require a carry to the next higher level of digit. If the next digit was already 9, this might generate another carry, and so on. At first, Babbage used the method of delayed sequential carry used in the Difference Engine. In this, the basic addition cycle was followed by a separate carry cycle. The carry cycle first performed any carry needed on the lowest digit, then proceeded to the next higher digit, and so on. This method worked, but it was slow because carries were performed separately for each digit. Babbage considered having 30 or 40 digits in each number column, so the carries might take a lot longer than the addition itself. Thus, a single multiplication might take some hundreds of separate addition steps. It was clear that the carry time had to be shortened.

Babbage tried various approaches to optimize the carries, and within a few months had adopted what he called the anticipating carriage. Additional hardware allowed the carriage mechanism to detect simultaneously where carries were needed and where one or more wheels already at 9 might cause a carry to propagate over a series of digits. All carries could be performed at once, regardless of the number of digits on an axis. Working out the details of anticipating carriage took Babbage many years, longer than any other single aspect of the machine. But it could speed operations greatly, justifying the effort. The mechanism was too complex to allow a carriage mechanism for each adding axis. Babbage was forced to adopt a design where a single anticipating carriage mechanism could be connected at will with any adding column through the central wheels.

Until then, multiplication had been provided by specialized hardware, and the carriage function had been removed from the adding axes to more specialized central hardware. Babbage soon realized that addition itself could be removed from the adding axes, and performed through the central wheels. The adding axes simply stored digits on their individual wheels, and they could be connected or disconnected from the central wheels as needed. Babbage separated the machine into a section of storage axes, which he called the store, and another section where operations were performed, which he called the mill. This very same division is found in all modern computers, though we now call them the memory and the central processing unit.

Simplification of the complex hardware for division led to new principles of control. In the first designs, elaborate mechanisms could sense the positions of the wheels holding the digits of the divisor and the remainder, then compare

The design drawing of the Analytical Engine shows the general arrangement of the machine, with circles usually representing a column of gears or wheels viewed from above.

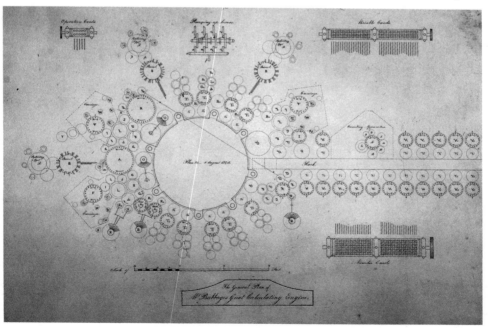

them one by one to determine if the next step should be a subtraction or a shift. Babbage later developed an elegantly simple alternative approach, in which the machine simply assumed that the divisor was smaller and performed the subtraction. If the assumption was wrong, the remainder would become negative. This would be evident from the topmost wheel on its axis, which indicated the sign. In this case, the machine would go into a special sequence where it would add the divisor back to the remainder, step the divisor down by one digit, and resume subtraction. This simplified the hardware considerably, and had major implications for general design. Babbage soon saw that he could use the change of sign on a numeric axis to control the flow of operations. In other words, if the result was positive, a mechanical test would cause the engine to continue one sequence of steps, but if the result was negative, the engine could switch to another sequence.

Not all changes were in the direction of simplification, however, for Babbage was very anxious to speed up calculation. An example of adding hardware to achieve speed was multiplication by table. In the initial method of multiplication by repeated addition, the number of cycles of addition would be equal to the sum of the digits of the multiplier. Thus, to multiply 38,471 by 694, for example, 38,471 would be added 19 times (6 + 9 + 4), along with 3 shifts. Babbage planned to work with numbers having as many as 40 digits. Multiplying two 40-digit numbers together might well take 200 addition cycles.

Babbage realized that by devoting a few cycles at the start of a long multiplication to some preparation, he could greatly speed the multiplication itself. He called this multiplication by table. In 9 cycles, he could calculate and place on special table axes in the mill the first 9 integral multiples of the multiplicand. Then he could simply pick one of these for each digit of the multiplier and add it to the accumulating product. Multiplying two 40-digit numbers would then take only 40 addition cycles, plus 9 to form the table, a total of 49 addition cycles rather than some 200. A similar method of division by table could also speed division.

By early 1835, the new engine had many different kinds of specialized axes that needed to be interconnected in a variety of ways, depending on what operation was under way. This required a more sophisticated approach to controlling the machine, and the one Babbage worked out over the next year involved several primary components.

The basic problem was one of controlling which axes would be connected to which central wheels at any given time. For this purpose Babbage devised a cylinder with studs on it, a "barrel" much like the cylinder in a mechanical music box where the tiny studs cause levers to strike various notes as the cylinder is rotated. The studs on Babbage's barrel were of different lengths, and as the barrel was rotated, one step at a time, the studs activated control levers that in turn set the positions of small gears (the "pinions"). These free-turning pinions would, in one position, connect the wheels on a chosen axis with the central wheels. In the other position, they would leave them unconnected. In this way, under the control of the studs on the cylindrical barrel, numbers could be transferred from one axis to another.

This general approach to control was an enduring aspect of the engine's design, but its function changed quite significantly. In the summer of 1835, the design called for each number axis and specialized function axis to have its own barrel, with permanently arranged rows of studs for each basic arithmetic operation. The rotational positions of each of these cylinders were controlled by the studs of a central drum, a similar cylinder with rows of studs that could be set by hand for a desired sequence of operations. Thus, one row of studs on the central drum might be set with studs that had the same effect as saying: "set the pinions so that variable axes 7 and 24 are fed to the multiplying columns." The next row might say "tell the multiplying columns to go into their multiplying sequence." And the next row might say "take the product indicated by the multiplying columns and store the result on variable axis 32."

This was a vastly sophisticated and flexible way to control a machine. But even without building experimental prototypes, Babbage realized that it was inadequate to control the calculating engine in the very complex work it would be capable of. The task of manually resetting studs in the central drum to tell the machine what to do was too cumbersome and error-prone to be reliable. Worse, the length of any set of instructions would be limited by the size of the drum.

His struggle with the problem of control led Babbage to a real breakthrough on June 30, 1836. He conceived of providing instructions and data to the engine not by turning number wheels and setting studs, but by means of punched card input. This did not render the central drum obsolete nor replace it. Punched cards provided a new top level of the control hierarchy that governed the positioning of the central drum. The central drum remained, but now with permanent sequences of instructions. It took on the function of micro-programming, so familiar to recent generations of computer engineers.

Babbage did not create the idea of punched cards out of thin air. Their use was widespread and well known in the control of cloth looms. This approach was invented in the 18th century by the Frenchman Jacques de Vaucanson and improved and commercialized around the turn of that century by his countryman Joseph Jacquard. In Jacquard's machines, a series of heavy pasteboard cards with holes in appropriate positions were strung together at their edges by a continuous ribbon. At any given step, a particular card would be pressed into a set of levers, which controlled the heddles, the wires that moved the warp threads to determine the pattern woven into the cloth. Jacquard's loom was the first to allow automatic control of elaborate patterns.

By 1900, punched cards had become a major device for tabulating numbers. They were introduced by Herman Hollerith, a statistician with the U.S. Bureau of the Census. He used them in the 1890 census to record data mechanically. By then, Hollerith could replace Babbage's mechanical rods with electrical brushes. Soon, machinery was devised for sorting and counting stacks of cards. This equipment provided the main input-output device for computers from 1945 until about 1980. Babbage's use of punched cards quite remarkably foreshadowed their later use, even though his version depended on mechanical sensing of holes rather than the electrical brushes of Hollerith and those who followed him.

If one were forced to chose some single date when the transition from the Difference Engine to the Analytical Engine was "complete," when the latter machine was finally "invented," it would probably be June 30, 1836, when punched cards were selected as the input mechanism. A

Instructions and data were entered into the Analytical Engine using punched cards. The smaller Operation Cards specified arithmetic operations to be performed; the larger Variable Cards dictated the "addresses" of the columns where the numbers to be operated on were found and where the results should be placed.

complete design would take another year and a half, until December 1837, when Babbage finally wrote an extended paper, "Of the Mathematical Powers of the Calculating Engine," which described the machine. Babbage continued design work for many more years, but this involved refinement of detail and alternatives of implementation, not changes of principle. By 1837, Babbage had devised a machine whose basic organization would remain unchanged through all his subsequent work, and indeed through the entire subsequent development of computer design.

The design principle comes down to a fourfold division of machine functions.

1) Input. From 1836 on, punched cards were the basic mechanism for feeding into the machine both numerical data and the instructions on how to manipulate them.

2) Memory. For Babbage this was basically the number axes in the store, though he also developed the idea of a hierarchical memory system using punched cards for additional intermediate results that could not fit in the store.

3) Central Processing Unit. For Babbage, this was the mill; like modern CPUs it provided for storing the numbers being operated upon most immediately (registers); hardware mechanisms for subjecting those numbers to the basic arithmetic operations; control mechanisms for translating the user-oriented instructions supplied from outside into detailed control of internal hardware; and synchronization mechanisms (a clock) to carry out detailed steps in a carefully timed sequence.

4) Output. Babbage's basic mechanism was always a printing apparatus, but he had also considered graphic output devices even before he adopted punched cards for output as well as input.

The way the store and the mill were organized and interconnected in Babbage's engine, as described in the 1837 paper, can be seen in the figure on page 85, which does not include input and output. The diagram is a plan view of the engine, as if looking down on it from above. At the top is a section of the store including six variable axes, labeled V1 through V6. In practice, the store would have been much longer, with many more variable axes; Babbage sometimes considered a minimum of 100, and as many as 1000. Each variable axis contained many figure wheels rotating around a central axle, each holding one digit of its variable. Babbage usually planned to have 40 digits per variable. One extra wheel on top recorded whether the value was positive or negative.

Running horizontally between the variable axes were the racks, long strips of metal with gear–toothed edges that carried digits back and forth between the store and the mill. Small movable pinions were positioned either to connect a given variable axis to the racks or to leave it unconnected. If a number was going into the mill, the racks would also be connected to the ingress axis in the mill (labeled I). From there, it would be passed to another appropriate part of the mill. When the mill was finished operating on a number, it would be placed on the egress axis (labeled E). This could then be connected to the racks, which would pass the number along to whatever variable axis had been chosen to hold the result.

The mill is the bottom section arranged around the large central wheels that interconnect its parts. For clarity, not all aspects of the engine are shown in this diagram. But this may obscure the machine's complexity and size. The central wheels alone were more than 2 feet across. The mill as a whole was about 4.5 feet in each direction. A store with 100 variable axes would have been about 10 feet long.

The ingress axis had its own anticipating carriage mechanism, labeled F1; an addition or subtraction could be performed there and then passed directly to the egress axis for storage. If a multiplication was coming up, the first nine multiples would be added on the ingress axis and stored on the table axes, shown as T1 through T9.

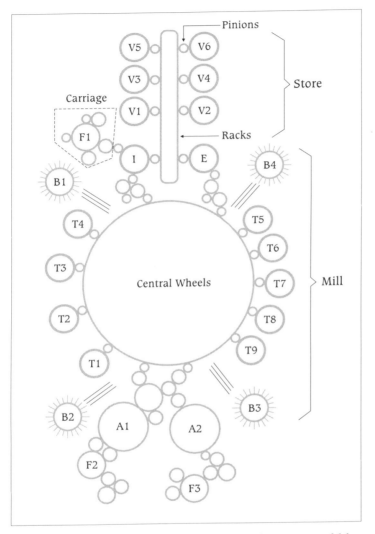

The results of a full multiplication or division would be formed on the two columns labeled A1 and A2 at the bottom of the diagram. This made it possible to hold intermediate results in "double precision" form. That is, if two 40-digit numbers were multiplied together, 80 digits of result could be kept on axes A1 and A2. A subsequent division by another 40-digit number still allowed 40 digits of precision in the result. A short example, using three-digit numbers, will make this clear. When 111 is multiplied by

999, the result is a six-digit answer, 110,889. If the result is now divided by 222, the first three digits of the answer are correctly 494. If, however, only the three highest digits had been retained from the multiplication, 110, then a slightly erroneous answer, 495, results from the division that follows. Modern computers retain a "double precision" step for multiplication to ensure the precision of the results of these arithmetic operations.

Axes F2 and F3 show the anticipating carriage mechanism for the A columns. Shown in symbolic form (B1 through B4) are four of the barrels with projecting studs that controlled the internal operations of the mill.

Some further details of control are shown in a simplified form in Figure M on page 87. This is a cross section, as if one were peering through the racks from one end of the machine. One sample variable axis is shown toward the top left; only 4 out of 40 figure wheels are shown in this diagram. In this case, the pinion axis P1 is in a lowered position, leaving the variable axis disconnected from the racks. It could be connected, though, if so ordered by a variable card. A series of these, strung together (edge on), can be seen at the left of the picture, hanging over the cardholder C. The arm A1 pivots about its center, and a small rod R1, projecting from its top, senses whether there is a hole in the current variable card corresponding to the variable axis shown. If there is a hole, the top of A1 can pivot to the left and the bottom to the right. That action would cause the slide S1 to lock into the flange shown on the bottom of the P1 pinion axis. Now, TP1 is part of a traveling platform, which moves up and down a small amount at appropriate times, carrying S1 with it. If, when it rises, S1 is locked into P1, P1 will be carried up as well, connecting the wheels of the variable axis with the corresponding racks.

To the right of the racks are shown a few wheels from a mill column adjacent to the racks (either the ingress or the egress axis). As shown, its wheels are connected to the racks

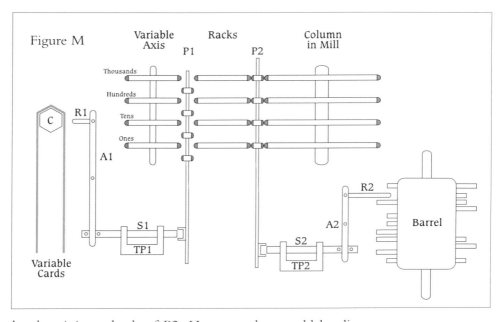

Figure M

by the pinion wheels of P2. However, they could be disconnected, if directed by the barrel shown at the lower right. The arm A2 pivots about its center. The small rod R2 on its top might be pushed leftward by an adjacent stud in the row down the left side of the barrel. But none is there, so R2 could pivot to the right, causing slide S2 to move to the left, and lock into the flange on the bottom of column P2. If the traveling platform TP2 then went through a downward motion cycle, P2 would also descend, disconnecting the racks from the mill column.

The barrel shown in the illustration has only about 10 stud positions in each vertical row. In the actual machine, the barrels were much larger because they controlled and coordinated the interaction of thousands of parts. Each row could contain as many as 200 stud positions, and each barrel could have 50 to 100 separate rows. The overall machine had several different barrels controlling different sections. Naturally, the barrels had to be closely coordinated with one another. In normal operation, each barrel would be rotated from one row to the adjacent one for successive machine cycles. Special events would determine some other treatment. For example, the barrels themselves could instruct a given row to turn back by a certain number of rows so a sequence could be repeated.

Figure M shows only variable cards as an external form of input. In practice, there were four different kinds of punched cards with different functions. Number cards were used to specify the value of numbers to be entered into the store, or to receive numbers back from the store for external storage. Variable cards specified which axes in the store should be the source of data fed into the mill or the recipient of data returned from it. In modern parlance, they supplied the memory address of the variables to be used.

Operation cards determined the mathematical functions to be performed. The logical content of an operation card might have been like this example: "Take the numbers from the variable axes specified by the next two variable cards, and multiply them in the mill; store the result on the variable axis specified by the third variable card." This was interpreted by the sensing rods on the operation-card reading apparatus and internally translated like this: "Advance the variable cards by one position, and rotate all the barrels to the starting position for a normal multiply-and-store sequence." Combinatorial cards controlled how variable cards and operation cards turned backward or forward after specific operations were complete. Thus, an operation card might have a logical content like this: "Move the variable cards ahead 25 positions, and set the operation cards to the start of the set that tells how to extract a square root."

Babbage planned to intersperse the combinatorial cards with the operation cards they controlled, so the four sets of cards required only three card readers (plus one card punch, for number cards being output from the machine). Thus, Babbage had devised a machine at least as sophisticated in its control principles as those that followed it until around 1950. Small details of operations were determined by the basic hardware. The hardware was controlled by the barrels. The barrels were controlled by the variable and operation cards, and these in turn were controlled by the combinatorial cards.

Some recent historians of computing regard the three-fold division of the instruction input into variable, operation, and combinatorial cards as archaic. We are used to a process where control-flow, operations, data, and storage addresses are specified in a single "instruction stream." But this underestimates what Babbage accomplished in the way of designing a programming interface despite never having an actual machine that he needed to program.

It would be very fair to say that the series of operation cards provided not a program, in current terms, but a series of subroutines. The combinatorial cards provided terminology, a control-flow program, invoking subroutines with call-by-reference values provided by the variable cards. Babbage's programming concepts clearly included what we call loops, subroutines, and branches (what later generations of programmers called "if" or "if-then" instructions).

Babbage realized that programs or subroutines (certainly not terms that he used) would need to be verified—what we would call "debugged." He also knew that it would be valuable to rerun verified programs on new sets of data, and even to share programs across multiple engines. Thus, it was a natural and practical approach to specify the data as being independent of the operations. Since he had no experience in programming an actual computer, it is not surprising that Babbage did not get to the modern concepts of higher level languages, interpreters, or compilers. His ideas for what amounts to assembly language programming do not correspond precisely to what emerged in the early 1950s, at the start of the electronic computer era. Nevertheless, there is no reason to suggest that his conceptualization was in any way basically inferior. It was in some ways more abstract, more powerful, and more general.

Although Babbage did not really expect to finish a single Analytical Engine, he knew that engine design would evolve over time. This meant that his initial card format might also change. Thus, he anticipated the need for specialized machines that could convert between the formats used by different models of Analytical Engines. Babbage knew that his machine would *in theory* make possible far more extended and precise calculations than had ever been attempted by hand. They would be possible *in practice* only with a machine that was highly reliable and quite fast. From his earlier work, he knew that reliability required the gears not to turn too quickly. Overall speed had to be achieved with smart design rather than raw power. This is what motivated the immense ingenuity that Babbage invested in time-saving methods like anticipating carriage and multiplication by table.

In the machine design of the late 1830s, the isolated addition of two 40-digit numbers would have taken about 19 seconds. But a lot of this involved moving numbers around between different sections before or after the actual addition. Babbage figured out how to overlap the different parts of the operation when more than two additions were to be performed in succession. This meant that each extra 40-digit addition took only 3.1 seconds.

Multiplication and division were similarly accelerated by clever logical design. The duration depended on the number of digits in the numbers. Take the case of a multiplication of 20 digits by 40 digits (a very high degree of precision even by current standards). With sustained additions at 3.1 seconds each, a straightforward step and add approach would have taken nearly 8 minutes to complete. Babbage was able to reduce this to less than 2 minutes. Today, with microprocessor speed measured in millions of multiplications per second, 2 minutes seems incredibly slow. But it was a remarkable accomplishment more than a century before electronic computation.

Finally, it is interesting from today's perspective to note that Babbage realized machine speed might well be affected by the numeric base in which variables were represented. Babbage's designs started out assuming our familiar decimal system, using base 10. But he also considered other internal representations, with bases ranging from binary (base 2) to centesimal (base 100). Never establishing that this would improve performance, he always returned to a decimal design. Modern computers are all essentially binary because the electronic devices from which they are built have only two basic states: on and off. This is not true of gear wheels. Using binary representation in a mechanical computer would have greatly increased the number of parts and over-all size and made the engine much slower. Babbage stuck with a decimal machine as a deliberate design choice.

6

Passages in a Philosopher's Life

In 1861, at the age of 70, Charles Babbage became more aware of his own mortality. He began to devote part of his time to writing a collection of reminiscences. Titled *Passages from the Life of a Philosopher*, it was published in 1864. Organized by topics rather than chronologically, this work provides many of the anecdotes that we have used to illuminate his life. Babbage called himself a philosopher because his activities ranged far beyond the narrow confines of the mathematics that had inspired his youthful studies. Natural philosophy was the term used to describe his interests in astronomy, physics, geology, and chemistry. Only in 1840 did a tutor at Trinity College, Cambridge, suggest the term "scientist" to identify people engaged in scientific activities. His name was William Whewell.

Becoming Master of Trinity College in the following year, Whewell frequently crossed paths with Babbage, and occasionally (in a metaphorical sense) they crossed swords. In the 1830s, Whewell and seven other men were invited to write books to show that the power, wisdom, and goodness of God are demonstrated throughout the whole of creation. Their publication was funded by a bequest of £8000, left

This photograph was taken at the Fourth International Statistical Congress in London in 1860, when Babbage was 68.

by the Earl of Bridgewater to support the view that scientific findings reinforce religious beliefs. Whewell wrote the volume on astronomy and physics. Being a clergyman as well as a scientist, Whewell expressed his opinion that the work of mathematicians did not in fact contribute to deepening human understanding of God.

Babbage disagreed so strongly with this opinion that he wrote a book to dispute it. Calling it the *Ninth Bridgewater Treatise*, he published it at his own expense in 1837. Taking up such issues as miracles and the age of Earth, Babbage used mathematical reasoning to reach the same conclusions that others derived from the Bible. He would not allow his former friend, Whewell, to get away with any slighting of the value of mathematics.

Another former friend was George Biddell Airy, who became England's Astronomer Royal in 1835. Though younger than Babbage, Airy had preceded him in the Lucasian professorship at Cambridge. However, Airy resigned that chair after only two years in order to take a much more lucrative post as professor of astronomy. Both Airy and Whewell had more support from the government than Babbage—they took the conservative positions that most prime ministers of the time espoused. As Astronomer Royal, Airy became the chief scientific adviser to the government. He was appointed to a government commission to inquire into railway gauges. In that role, Airy staunchly opposed Babbage and his support of Brunel's 7-foot gauge.

In the British Association, Babbage promoted contacts between scientists and the industrialists in each locality where the annual meeting was held. He proposed that the BA sponsor a display of industrial products at each meeting. This would also encourage scientists to get involved with industrial progress. Conservatives like Whewell and Airy opposed such contacts, and their influence carried the day. Babbage resigned his trusteeship of the BA in 1839. He was surely helped to that decision by a remark of Whewell's. To further his aims of industrial science, Babbage had joined with others in recommending that the BA meeting of 1837 be held in the great industrial center of Manchester. Someone pointed out that an attraction of Manchester was its statistical society. Whewell said that was a very good reason for *not* going to Manchester.

In 1851, Babbage found an occasion to vent his anger against these men. Through the 1840s, friends of the Prince Consort, Queen Victoria's husband, urged him to sponsor an international exhibition of industrial products to display Britain's superiority. The French had been holding such exhibitions every five years since 1800. The Great Exhibition was planned for 1851. With Britain finally

George Biddell Airy, an English astronomer, was a great rival of Babbage. Airy contributed to studies of light and of the earth's motions.

bestirring herself, Babbage naturally expected to be consulted for his interest and experience. In fact, his friend Lyon Playfair, one of the exhibition commissioners, proposed his name. However, government authorities had no wish to deal further with the old scientific radical.

In his typical fashion, Babbage decided to give his advice anyway. He published a book of 200 pages, *The Exposition of 1851: Views of the Science and the Government of England*. Babbage made sensible recommendations about the siting of the exhibition hall and suggested that trams within the building, 1600 feet long, would make it easier for the public to view the exhibits. He also devoted a number of pages to a stinging criticism of the Astronomer Royal, George Airy. This man, he wrote, "wishes himself to be considered the general referee of government in all scientific questions." Babbage pointed out that Airy was spreading himself so thin that he was shirking his major responsibilities for the Greenwich Observatory and the *Nautical Almanac*.

The Great Exhibition was housed in a brilliant architectural structure by the engineer Joseph Paxton. Originally a gardener, Paxton designed a gigantic greenhouse. Built of iron and glass, it was soon dubbed the Crystal Palace. It was erected in Hyde Park in the brief space of seven months. After the exhibition, it was disassembled and moved to a permanent site just south of London. Airy opposed the original construction, saying that its flimsy structure would collapse in a strong wind. He was wrong. Thirty years later, Airy would be disastrously wrong in the opposite direction. During the building of a bridge over the Firth of Tay in Scotland, the engineer consulted Airy on the wind load to be expected. Airy gave his opinion as 10 pounds per square foot. Designed to that specification, the bridge collapsed within a year. The next engineer made his own measurements, calculated wind loads of 34 pounds per square foot, and designed his structure to withstand 56. Airy held the post of Astronomer Royal, worth more than £1300 per

The Crystal Palace was built for London's first Great Exhibition of Manufactures in 1851. Exhibitors from Great Britain and abroad numbered close to 14,000.

year, for 45 years, relinquishing it only at age 80. Babbage felt that Airy's tenure had done little to enhance the prestige of science in England.

Before the Great Exhibition, Babbage and Airy had already clashed seriously in the council of the Royal Astronomical Society. The issue involved the awarding of medals to honor the discovery of the planet Neptune. That discovery was a great triumph for Newton's theory of gravitation because two mathematicians had used it to predict Neptune's location before any observer identified it. The few observations of Uranus made since William Herschel's discovery 60 years earlier did not fit a uniform orbital path about the sun. A Cambridge student, John Adams, age 26, calculated the position of an unknown planet that could disturb the orbit of Uranus. In the fall of 1845, Adams sent his calculations to Airy, who paid little attention to them. Eight months later, the French astronomer U. J. J. Le Verrier published essentially the same results, and sent a copy to Airy. Airy responded immediately, without mentioning Adams.

SOCIÉTÉ FRANÇAISE
de Statistique Universelle.

LE ROI, PROTECTEUR.

La Société a été fondée à Paris, par M. CÉSAR MOREAU, de Marseille le 22 Novembre 1829.

DIPLOME DE MEMBRE HONORAIRE

de Monsieur Charles Babbage, Membre de la Société Royale de Londres &c. &c. &c. &c. &c. &c.

La Société Française de Statistique Universelle, d'après le Rapport qui lui a été fait par son Conseil d'Administration, a admis au nombre de ses Membres Honoraires la personne sus-dénommée. En foi de quoi nous lui avons fait expédier le présent Diplôme.

Paris (Bureaux de la Société, rue Vivienne, 5 10) le 22 Juillet 1834.

Babbage was highly critical of the state of science in England and particularly the conduct of learned societies. This contrasts with his high regard for the scientific academies of France, Italy, and Prussia. This certificate made Babbage an honorary member of the Société Française de Statistique Universelle.

Then, while Airy set slow wheels in motion to search for Neptune, Le Verrier also sent his results to J. G. Galle at the observatory in Berlin. Galle received the letter in September 1846, and identified Neptune the same night.

At council meetings of the Royal Astronomical Society early in 1847, Airy, supported by Whewell, opposed the awarding of a society medal to Le Verrier. Babbage believed that Adams and Le Verrier had made equally eminent contributions. He proposed giving the 1846 medal to Le Verrier (who had the priority of publication), and the 1847 medal to Adams (who had the priority of invention). However, the council finally adopted the unhappy compromise of awarding no medal for 1847. In 1848, instead of a medal, they gave testimonials to 12 men, some of whom had made rather minor contributions. The 12 included Airy, Adams, and Le Verrier, but not Galle. The final comical twist of this sordid episode is that Airy had already been awarded the 1846 medal for doing his job as Astronomer

Royal. Clearly, Babbage believed that Airy had not been doing his job.

With his restless mind, Charles Babbage was forever using his ingenuity to benefit society. In 1851, he conceived a way to enhance safety at sea. For coastal navigation, captains often used lights on shore to help fix their position. In some areas, lights were numerous enough to confuse the captains. Babbage proposed to control the emission of lights, such as lighthouses and harbor markers, in a way that would identify each light. He wanted each light to flash intermittently so it would broadcast a unique number. In his typical fashion, Babbage devised a mechanism for such signaling. One paragraph from the report he wrote shows that he was applying principles gained from designs of his Difference Engine. His basic idea was to enclose the light in a hollow cylinder with a hole in it. Raising and lowering the cylinder would alternately block and expose the light. Babbage wrote:

> . . . great accuracy in the [controlling] wheelwork is necessary. In lighthouses the moving power may be a heavy weight driving a train of wheels. This must terminate in a governor, which presses by springs against the inner side of the cylinder. . . . The governor must be so adjusted that some one axis shall revolve in the given time. A cam-wheel must be fixed on this axis, having its cams and blank spaces so arranged as to lift up the tail of a lever carrying the occulting cylinder at the proper intervals of time. Each tooth of the cam-wheel will cause an occultation of the lamp by the cylinder, which is instantly drawn back by a spring.

As you might expect, Babbage built a model of this system, and displayed it from an upper window of his house. He sent his report on occulting lights to a number of governments. The English corporation responsible for lights and buoys apparently did not respond. However, when he demonstrated a model in Brussels in 1853, a Russian naval officer showed great interest. The Russians used the principle during the Crimean War against Britain and France. The United States was also interested; Congress granted $5000 to investigate Babbage's system. He was invited by an American representative to return with him to assist in the experiments. Although Babbage was sorely tempted, he felt the press of other work too strongly, and declined. He saved his life because the ship to America collided with another off the coast of Newfoundland. His friend and many other passengers perished.

Until the early 1850s, Babbage remained on friendly terms with Ada and Lord Lovelace, although collaboration on the Analytical Engine had ceased. Ada turned to more dangerous activities and contracted severe debts from betting on horse races. And by 1850, she was seriously ill with uterine cancer. Ada turned to Babbage for financial advice. He did what he could, but she was dominated by her remorseless mother. Ada died in 1852 at age 36. Her mother's disputing of Ada's will caused great bitterness and ended Charles's relations with the Lovelace family.

More happily for Babbage, his son Henry returned from India on a three-year furlough in 1854. Charles welcomed him and his wife warmly and built a comfortable nursery for their child in his home. As with many men, Charles Babbage proved to be a more loving grandfather than he had been a father. At the same time, he and Henry became good friends. They attended parties together and traveled around England. Henry also engaged in mathematical studies to assist his father.

Just at that time, a Swedish engineer brought a calculating engine to England. In 1834, George Scheutz had read an article describing Babbage's Difference Engine. He resolved to make one for himself. He and his son worked for many years to design and build a working engine. With occasional support from the Swedish government, they eventually achieved success, using a number of mechanical principles that were quite different from Babbage's. The

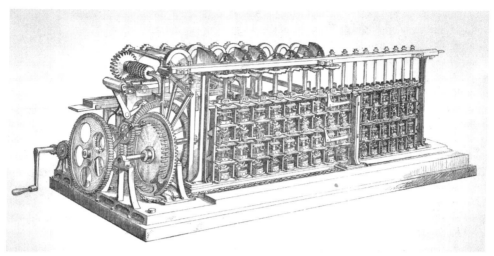

Scheutzes were apprehensive about how Babbage would view their competition. They need not have worried—Charles supported them enthusiastically. He helped to ensure that their engine won the gold medal of the Paris Exposition of 1855. In England, Henry made two large plans of the Scheutz engine using his father's mechanical notation—one of the them was 13 feet long by 3 feet wide. At the 1855 Edinburgh meeting of the British Association, Henry Babbage gave a lecture on the mechanical notation and the workings of the Scheutz engine.

Charles's friend, the engineer Bryan Donkin, made a replica of the machine, which was later used in England to print some mathematical tables for the government. The original Scheutz machine was purchased by an American, who put it in the Dudley Observatory in Albany, New York. As a final proof of his great good will toward the Scheutzes, Babbage proposed to the Royal Society in 1856 that George should be awarded one of its royal medals—to no avail.

At the end of 1856, Henry's furlough ended. Charles bid his family a fond farewell and returned sadly to his empty house. Their happy interlude was over. Charles turned back to his Analytical Engine, to which he made

A barefoot newsboy holds a poster for the Pall Mall Gazette of October 21, 1871, which carried the news of Babbage's death.

significant further improvements. Though never constructed, his final designs incorporated principles that brought him ever closer to the concepts of the general-purpose computer that would be reinvented in the 1940s and 1950s.

Babbage's friend and first biographer, Harry Buxton, wrote of his mentor that he "sought and cultivated the society of educated women, in whose elegant accomplishments and lively conversation he endeavored to temper the severe studies of his ordinary pursuits." In his later years, Babbage corresponded with Jane Harley Teleki, daughter of one of his old friends. When she was in Turin briefly in 1863, Jane sent a note from Charles to Luigi Menabrea, by then a prominent government official. He immediately left his office to spend the evening with this friend of his respected Babbage.

We get a touching glimpse of Babbage in his old age from a letter he wrote to Jane:

> My lonely household has been relieved of some of its dreariness by the arrival of a fair young creature who gives me a joyous greeting every morning at my breakfast table. She sits quietly by my side whilst I am working in the drawing room, and in the evening delicately reminds me that it is time to retire to rest by saying "Polly wants to go to bed," on which I ring the bell and the servant covers up her cage with a curtain whilst I dream of another far away.

A mutual friend of Jane and Charles was Margaret, Duchess of Somerset. She was the second wife and now widow of Charles's longtime friend, Edward Seymour,

Duke of Somerset. Charles and Margaret enjoyed each other's company during their later years. On one occasion, inviting him to dinner, she wrote: "Pray come and meet the Turkish Ambassador & the Spanish Ambassador just arrived & his charming lady—and be here at Dinner *next Thursday* 5th of Oct at 8 o'clock and assist at a *magnifique haunch de venison* sent by our excellent friend Lord Dalhousie. Pray send a favorable answer to yours ever . . ."

Charles Babbage died on October 18, 1871, just short of his 80th birthday. At the end, he was assisted by his son Henry, again on furlough, and his brother-in-law and schoolmate, Edward Ryan. As family and a few friends walked his casket to the cemetery, they were accompanied by one carriage, belonging to the Duchess of Somerset.

A fitting epitaph was written by Joseph Henry, director of the Smithsonian Institution of Washington, who had visited Babbage in 1870:

> Hundreds of mechanical appliances in the factories and workshops of Europe and America, scores of ingenious expedients in mining and architecture, the construction of bridges and boring of tunnels, and a world of tools by which labor is benefited and the arts improved—all the overflowings of a mind so rich that its very waste became valuable to utilize—came from Charles Babbage. He more, perhaps, than any man who ever lived, narrowed the chasm [between] science and practical mechanics.

After Babbage

THE BABBAGE ENGINE PROJECT

The Science Museum builds a 19th century 'computer'

In 1821, Charles Babbage dreamed of producing mathematical tables mechanically. When he died 50 years later, no Babbage engine was in operation producing tables on a regular basis. He had not realized his dreams. But it would be wrong to write off those dreams as a failure.

Babbage was not an impractical technologist reaching beyond his grasp, but a creative scientist exploring ways to realize mathematical relations in mechanical form. As we have seen, many of the sections and components of his engines were in fact very ingenious solutions to difficult problems. As it has turned out, the final complete fulfillment of the Babbage dreams depended on electrical and electronic mechanisms unavailable to him. He himself recognized the nature of the situation. He wrote in his *Passages from the Life of a Philosopher*:

> The great principles on which the Analytical Engine rests have been examined, admitted, recorded, and demonstrated. The mechanism itself has now been reduced to unexpected simplicity. Half a century may probably elapse before anyone without those aids which I leave behind me, will attempt so unpromising a task. If, unwarned by my

Engineers Barrie Holloway (left) and Reg Crick (right) built the calculating part of Babbage's Difference Engine No. 2 in 1991 as part of the Science Museum of London's commemorative Babbage Engine Project.

example, any man shall undertake and shall succeed in really constructing an engine embodying in itself the whole of the executive department of mathematical analysis upon different principles or by simpler mechanical means, I have no fear of leaving my reputation in his charge, for he alone will be fully able to appreciate the nature of my efforts and the value of their results.

Although lacking foresight into the turns technology would take, Babbage did not err in his "half a century" by more than about 25 years.

The residue of Babbage's drawings and mechanisms were left to his son Henry. When he retired from his service in India, Henry Babbage published in 1889 a collection of articles and notes of his father's, along with additions of his own, with the title *Babbage's Calculating Engines*. He also used parts that had already been fabricated to construct six small demonstration models of the Difference Engine. He distributed them to universities in several different countries. In themselves, these efforts produced no signficant further work. Those familiar with the work seemed to think that "If Babbage couldn't do it, it can't be done."

Mechanical calculating proceeded gradually. De Colmar's arithmometer achieved commercial success after being exhibited at the Paris Exposition of 1867. In 1885, in the United States, William Burroughs introduced the printing adding machine that formed the basis of cash registers and other calculators. At first driven by a hand crank, these machines eventually operated with electric motors. The punched-card tabulators of Herman Hollerith gained influence in the early 1900s. In 1924, Hollerith's company merged with others to form International Business Machines (IBM).

Punched-card tabulating equipment evolved considerably in the first half of the twentieth century, and so too did arithmetic machines. Both were intended primarily for commercial applications but could be adopted as well for scientific computation. In both England and the United

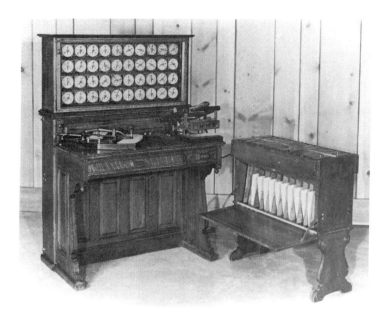

The Hollerith tabulator read data from holes in punched cards using tiny electrical switches. In 1924, Herman Hollerith's company merged with others to form International Business Machines (IBM).

States, commercial machines were modified to act as Difference Engines and used to compute mathematical and astronomical tables. Researchers in the United States and Germany began experimenting with electronic computing devices rather than simply electromechanical ones.

The first proposal to build a complete machine of similar complexity to Babbage's was made in 1937 when a physicist at Harvard University, Howard Aiken, conceived of a programmable electromechanical calculating machine. Aiken managed to interest the U.S. Navy in supporting his machine; IBM designed and built it. It was finished in 1944 and is often called the Mark I computer. It was 51 feet long, greater in size, though not in capability, than anything Babbage had proposed. Like Babbage's Difference Engine, the Mark I was designed and used primarily for calculating and printing mathematical tables. It could not be programmed nearly as flexibly as the Analytical Engine.

After Aiken's machine was started, but before its completion, World War II broke out. This greatly accelerated the application of electronics to a wide range of practical and computational problems. Radar is certainly a prime

The Mark I was an early electro-mechanical computer, built at Harvard University in 1944 under the direction of Howard Aiken.

example. Another example, quite secret until recently, was the Colossus machine designed and built in England to decode messages processed by German cipher machines. A main figure in its design was Alan Turing, who was also a pioneer in the development of computer theory.

None of these machines was a general-purpose programmable computer in the modern sense. The first machine with any pretense at that title is the ENIAC (Electronic Numerator Integrator and Calculator) developed at the University of Pennsylvania between 1943 and 1945. Like the Mark I, it was initially intended to calculate tables, in its case, artillery tables for use in firing various kinds of weapons. But it was too difficult to program and had too small a memory to be a generally useful machine.

However, the general-purpose computer did appear in the very next generation of machines. With a computer that can in principle imitate any other, designers had finally caught up with the concept that Babbage had clearly formulated more than a century before. Four examples are the EDVAC, a follow-on to the ENIAC; the Whirlwind, devel-

oped at the Massachusetts Institute of Technology; the EDSAC developed at Cambridge University and the IAS Computer developed at the Institute for Advanced Studies in Princeton (New Jersey) under John von Neumann. From these projects flowed almost all later development of computers.

What influence did Babbage's work have on those who followed him and on the eventual emergence of the computer? The answer, while unsatisfying, is simple: it is not fully clear. Herman Hollerith did not derive punched-card tabulation from knowledge of the Analytical Engine, although he may have been familiar with the Jacquard loom. On the other hand, those who devised electromechanical Difference Engines in the twentieth century were certainly aware of Babbage's earlier work and considered themselves to be his technological heirs. Yet they probably did not derive details of their own machines from what he had done.

Howard Aiken clearly was aware of Babbage from a very early point in his own work, but the character of

Dr. J. Presper Eckert, Jr. demonstrates the ENIAC computer that he coinvented to calculate tables. ENIAC was much faster than Mark I, but was still too difficult to program and had too small a memory to be a generally useful machine like today's computers.

Howard Aiken (right) examines one of the typewriters of the Mark I as it performs its second-ever calculation. With Aiken are Robert V. D. Campbell (center), who supervised the construction of Mark I, and optician James Baker (left)

influence is hard to judge. It is unknown whether Aiken's first ideas of building a computing machine were inspired by knowledge of the Analytical Engine, or whether someone told him about it after his ideas had germinated. However, the first written proposal for the Mark I described both the Difference and Analytical Engines at some length. The first published book to describe the Mark I began with a lengthy account of Babbage's work, and praised the principles of the Analytical Engine's design. Indeed, when the book was reviewed in the leading British scientific journal, it appeared under the title "Babbage's Dream Comes True."

It is also clear, however, that Aiken was not strongly influenced by the details of Babbage's work. Aiken had access to Babbage's autobiography and to *Babbage's Calculating Engines*, which reprinted virtually everything that had been published on the Analytical Engine, including the Menabrea/Lovelace/Babbage paper. He could well have derived the idea of a general purpose programmable computing machine from these writings. If so, either he did not fully understand it or felt unable to implement it fully, for the programmability and flexibility of the *Mark I* were quite inferior to those of the Analytical Engine.

Further, the available published material described the Analytical Engine quite abstractly. No description of the actual machine or the many fundamental design choices with which Babbage had wrestled existed outside of his detailed drawings and engineering workbooks; and these were unavailable to Aiken. They were also unavailable to his successors because early in World War II they were all packed up in large crates and shipped from London to the remote countryside to protect them from destruction in urban bombing. They were not returned and made available until 1968.

The British Royal Mail issued this special postage stamp honoring Charles Babbage, who repeatedly expressed frustration at the lack of recognition for his work, in 1991.

So, the Analytical Engine had little or no direct influence on the engineering design of the computer when it finally emerged, though it either inspired or encouraged the general idea of a computer in Aiken's work. Was this latter role very important? Probably not. Aiken's work somewhat influenced what followed it, but the main impetus was the tremendous need for huge quantities of computation created by new weapons and new applications of science to warfare during World War II. These attracted many talented scientists, mathematicians, and engineers to the suddenly urgent problem of automatic computing, and their work would probably have proceeded as it did even if they had never heard of Babbage or Aiken.

Yet, those who built the first complete working computers recognized immediately that Babbage had, in principle, invented the same machine, and that while he cannot be credited with the engineering detail of electronic computers, he was very much their intellectual and spiritual ancestor and a heroic pioneer of the new computer era.

MUSEUMS AND WEB SITES RELATED TO CHARLES BABBAGE

American Computer Museum

234 East Babcock Street
Bozeman, MT 59715
Tel: 406-587-7545
http://www.compustory.com

The British Library

96 Euston Road
London NW1 2DB
United Kingdom
Tel.: 44-171-412-7332
http://www.bl.uk

Computer Museum of America

Coleman College
7380 Parkway Drive
La Mesa, CA 91942
Tel.: 619-465-8226
http://www.computer-museum.org

The Computer Museum

300 Congress Street
Boston, MA 02210
Tel.: 617-426-2800
Talking Computer: 617-423-6758
http://www.tcm.org

Deutsches Museum

Museumsinsel 1
D-80538 München
Germany
Tel: 49-89-2179-1
Fax: 49-89-2179-324
http://www.deutsches-museum.de

National Museum of American History

The Smithsonian Institution
14th Street and Constitution Ave., N.W.
Washington, DC 20560
Tel.: 202-357-2700 (voice)
        or 202-357-1729 (TTY)
Fax: 202-633-9338
http://www.si.edu
http://www.si.edu/organiza/museums/nmah

Science Museum, London

National Museum of Science and Industry
Exhibition Road
South Kensington
London SW7 2DD
United Kingdom
Recorded message: 44-171-938-8111
General Inquiries: 44-171-938-8008/8080
Disabled Persons Inquiry Line:
        44-171-938-9788
http://www.nmsi.ac.uk

Web Sites Related to Charles Babbage

The Babbage Foundation
http://www.scsn.net/users/babbage/index.html

Charles Babbage Institute at the
University of Minnesota
http://www.cbi.umn.edu

Personal information page on Charles
Babbage
http://www.comlab.ox.ac.uk/oucl/users/
jonathan.bowen/babbage.html

CHRONOLOGY

1791
Charles Babbage born, south London, December 26

1810–14
Attends Trinity College, Cambridge

1812–14
Member of the Analytical Society at Cambridge, which he helps found

1814
Marries Georgiana Whitmore in July

1815
First child, Benjamin Herschel Babbage, born

1815
Becomes a member of the Royal Society

1815–16
Publishes an essay on calculus in *Philosophical Transactions of the Royal Society*

1816
Presents series of lectures on astronomy at the Royal Institution in London

1819
Travels to Paris to visit French scientists; gets inspiration for Difference Engine from Baron Gaspard de Prony's use of division of labor for calculating tables

1820
Helps found the Astronomical Society of London

1822
Announces invention of Difference Engine to Astronomical Society in June

1823

Recognized by the Royal Society for his Difference Engine

1824

Is awarded the Astronomical Society's first gold medal

1826

Publishes *A Comparative View of the Various Institutions for the Assurance of Lives*

1826

Publishes description of his mechanical notation in *Philosophical Transactions of the Royal Society*

1827

Father Benjamin, son Charles Jr., wife Georgiana, and a newborn son die

1827

Consults with Isambard Kingdom Brunel, who is overseeing his father's tunnel under the Thames River, on railroad design

1827

Begins scientific tour of Europe with mechanic Richard Wright

1829–39

Lucasian professor of mathematics at Cambridge University

1830

Publishes *Reflections on the Decline of Science in England, and on Some of its Causes*

1831–39

Trustee of the British Association for the Advancement of Science

1832

Construction work on the Difference Engine halts

1832
Publishes *On the Economy of Machinery and Manufactures*

1834
Helps found the Statistical Society of London

1834
Daughter Georgiana dies

1836
First conceives of using punched cards to provide instructions and data to calculating machine—this marks transition from Difference Engine to Analytical Engine

1837
Writes "Of the Mathematical Powers of the Calculating Engine"

1837
Publishes *Ninth Bridgewater Treatise*

1843
Babbage and Ada Lovelace publish translation of Menabrea's description of the Analytical Engine

1844
Babbage's mother Betty dies

1851
The Great Exhibition, England's first exhibition of industrial products, is held; Babbage conceives of way to control the timing of the emission of light from lighthouses and harbor markers

1864
Publishes *Passages from the Life of a Philosopher*

1871
Charles Babbage dies on October 18

Asprey, William, ed. *Computing before Computers.* Des Moines: Iowa State University Press, 1990.

Atherton, W. A. *From Compass to Computer, A History of Electrical and Electronics Engineering.* San Francisco, Calif.: San Francisco Press, 1984.

Babbage, Charles. *Passages from the Life of a Philosopher.* New Brunswick, N.J.: Rutgers University Press, 1994.

Babbage, Henry Prevost. *Babbage's Calculating Engines: A Collection of Papers.* Los Angeles: Tomash, 1982.

Bell, Walter Lyle. *Charles Babbage, Philosopher, Reformer, Inventor: A History of His Contributions to Science.* Doctoral dissertation, Oregon State University. Ann Arbor: University of Michigan Microfilms, 1975.

Buxton, H. W. *Memoir of the Life and Labours of the Late Charles Babbage Esq., F.R.S.* (Anthony Hyman, ed.) Cambridge, Mass.: MIT Press, 1988.

Campbell-Kelly, Martin, ed. *The Works of Charles Babbage.* 11 vols. London: Pickering, 1989.

Cardwell, D. S. L. *Turning Points in Western Technology.* New York: Science History Publications, 1972.

Charles Babbage and His Calculating Engines. London: Science Museum, 1991.

Collier, Bruce. *The Little Engines That Could've: The Calculating Machines of Charles Babbage.* New York: Garland, 1991.

Dubbey, J. M. *The Mathematical Work of Charles Babbage.* Cambridge: Cambridge University Press, 1978.

Hyman, Anthony. *Charles Babbage: Pioneer of the Computer.* Princeton, N.J.: Princeton University Press, 1982.

Hyman, Anthony, ed. *Memoirs of the Life and Labours of the Late Charles Babbage, Esq.* Cambridge, Mass.: MIT Press, 1988.

Hyman, Anthony, ed. *Science and Reform: Selected Works of Charles Babbage.* Cambridge: Cambridge University Press, 1988.

Lindgren, Michael. *Glory and Failure: The Difference Engines of Johann Müller, Charles Babbage and Georg and Edvard Scheutz.* (Craig G. McKay, trans.) Cambridge: MIT Press, 1990.

MacLachlan, James. *Children of Prometheus: A History of Science and Technology.* Toronto: Wall & Emerson, 1990.

Moore, Doris Langley. *Ada, Countess of Lovelace: Byron's Legitimate Daughter.* London: John Murray, 1977.

Morrison, Philip, and Emily Morrison, eds. *Charles Babbage: On the Principles and Development of the Calculator and Other Seminal Writings.* New York: Dover, 1961.

Moseley, Maboth. *Irascible Genius: A Life of Charles Babbage, Inventor.* London: Hutchinson & Co., 1964.

Stein, Dorothy. *Ada: A Life and a Legacy.* Cambridge, Mass.: MIT Press, 1985.

Swade, Doron. *Charles Babbage and His Calculating Engines.* London: Science Museum, 1991.

Zientara, Marguerite. *History of Computing: A Biographical Portrait of the Visionaries Who Shaped the Destiny of the Computer Industry.* Framingham, Mass.: CW Communications, 1981.

Bruce Collier is a former assistant dean of Harvard College and a former principal engineer for Digital Equipment Corporation in Maynard, Massachusetts. He graduated *cum laude* from St. John's College and received an M.A. and a Ph.D. in the history of science from Harvard University. He is the author of *The Little Engines That Could've: The Calculating Machines of Charles Babbage.*

James MacLachlan, emeritus professor of history at Ryerson Polytechnic University in Toronto, is a freelance author and editor. He is the author of *Children of Prometheus: A History of Science and Technology* and *Galileo Galilei: First Physicist* (Oxford University Press, 1998). He is also the principal author of *Matter and Energy: Foundations of Modern Physics.*

Owen Gingerich is Professor of Astronomy and of the History of Science at the Harvard-Smithsonian Center for Astrophysics in Cambridge, Massachusetts. The author of more than 400 articles and reviews, he has also written *The Great Copernicus Chase and Other Adventures in Astronomical History* and *The Eye of Heaven: Ptolomy, Copernicus, Kepler.*